Interactive Reader and Study Guide

Holt Social Studies

Eastern World

HOLT, RINEHART AND WINSTON

A Harcourt Education Company

Orlando • **Austin** • New York • San Diego • London

How to Use this Book

The *Interactive Reader and Study Guide* was developed to help you get
the most from your world geography course. Using this book will help
you master the content of the course while developing your reading and
vocabulary skills. Reviewing the next few pages before getting started
will make you aware of the many useful features of this book.

*Chapter Summary pages help you connect with the big picture. Studying
them will keep you focused on the information you will need to be successful
on your exams.*

Graphic organizers help you to
summarize each chapter.
- Some have blanks you will need
  to fill.
- Others have been completed
  for you.
Either way, they are a valuable
study tool to help you prepare for
important tests.

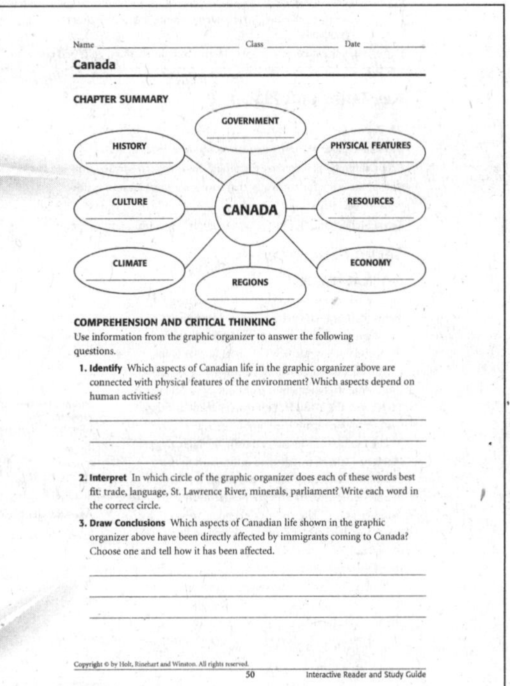

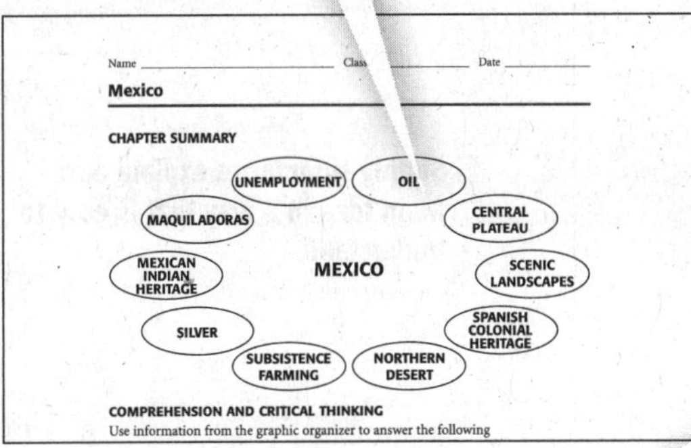

Answering each question will help
you to understand the graphic
organizer and ensure that you
fully comprehend the content from
the chapter.

Section Summary pages allow you to interact easily with the content and Key Terms from each section.

Main Ideas statements from your textbook focus your attention as you read the summaries.

Clearly labeled page headers make navigating the book extraordinarily simple.

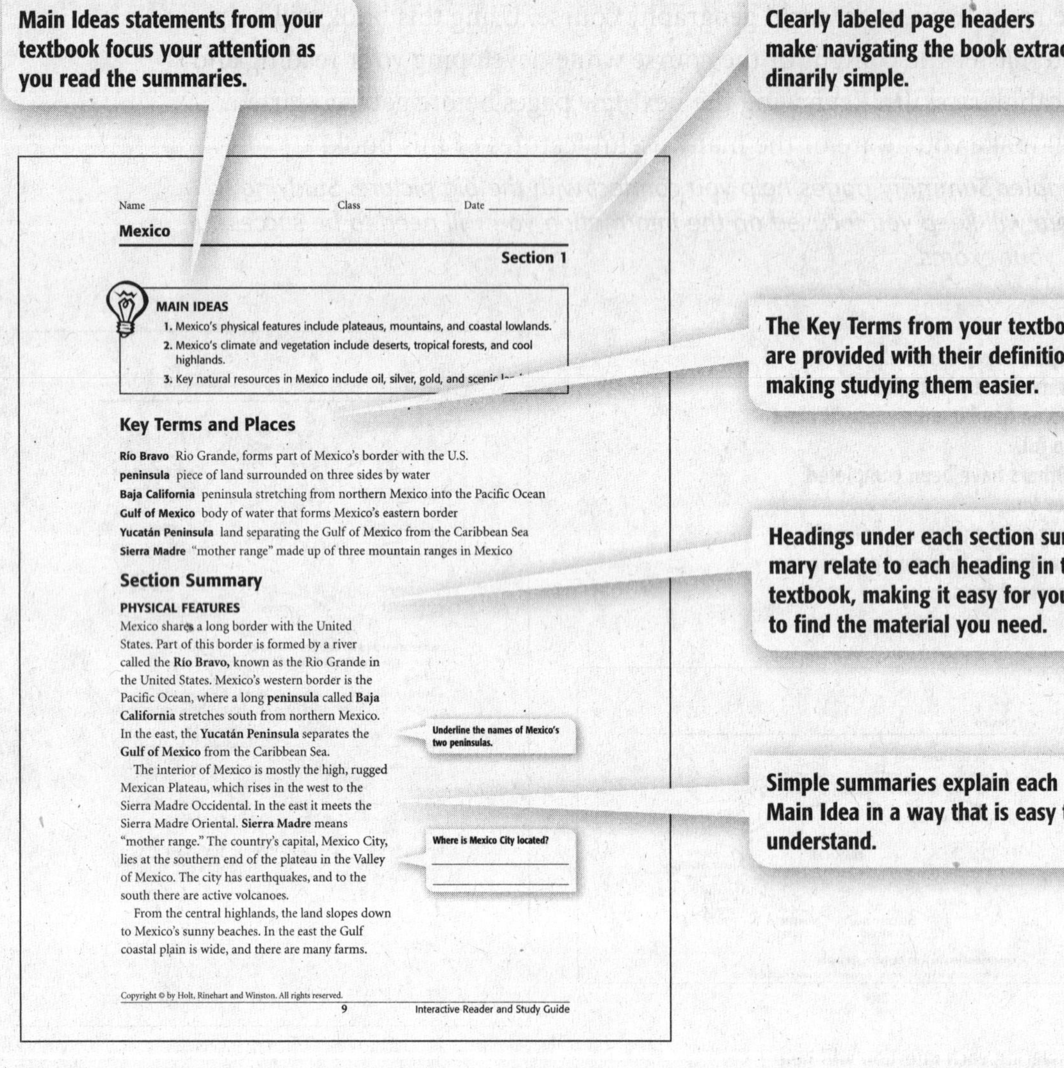

The Key Terms from your textbook are provided with their definitions, making studying them easier.

Headings under each section summary relate to each heading in the textbook, making it easy for you to find the material you need.

Simple summaries explain each Main Idea in a way that is easy to understand.

Notes throughout the margins help you to interact with the content and understand the information you are reading.

Challenge Activities following each section will bring the material to life and further develop your analysis skills.

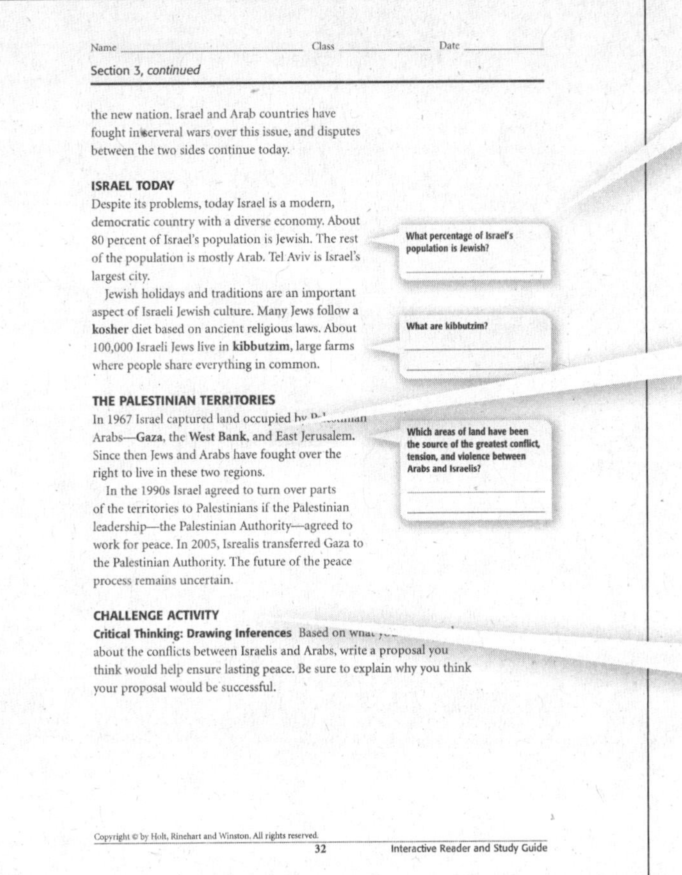

Name _____________ Class _____________ Date _____________

Section 3, *continued*

the new nation. Israel and Arab countries have fought in several wars over this issue, and disputes between the two sides continue today.

ISRAEL TODAY

Despite its problems, today Israel is a modern, democratic country with a diverse economy. About 80 percent of Israel's population is Jewish. The rest of the population is mostly Arab. Tel Aviv is Israel's largest city.

Jewish holidays and traditions are an important aspect of Israeli Jewish culture. Many Jews follow a **kosher** diet based on ancient religious laws. About 100,000 Israeli Jews live in **kibbutzim**, large farms where people share everything in common.

THE PALESTINIAN TERRITORIES

In 1967 Israel captured land occupied by Palestinian Arabs—**Gaza**, the **West Bank**, and East Jerusalem. Since then Jews and Arabs have fought over the right to live in these two regions.

In the 1990s Israel agreed to turn over parts of the territories to Palestinians if the Palestinian leadership—the Palestinian Authority—agreed to work for peace. In 2005, Isrealis transferred Gaza to the Palestinian Authority. The future of the peace process remains uncertain.

CHALLENGE ACTIVITY

Critical Thinking: Drawing Inferences Based on what you about the conflicts between Israelis and Arabs, write a proposal you think would help ensure lasting peace. Be sure to explain why you think your proposal would be successful.

What percentage of Israel's population is Jewish?

What are kibbutzim?

Which areas of land have been the source of the greatest conflict, tension, and violence between Arabs and Israelis?

32   Interactive Reader and Study Guide

Be sure to read all notes and answer all of the questions in the margins. They have been written specifically to help you keep track of important information. Your answers to these questions will help you to study for your tests.

The Key Terms from your textbook have been boldfaced, allowing you to find and study them quickly.

Challenge Activities ask you to think critically about the material from the section. Completing each of these exercises will increase your understanding of the material and develop your analysis skills.

A Geographer's World

CHAPTER SUMMARY

<table>
<tr><td colspan="3" align="center">GEOGRAPHY</td></tr>
<tr><td>Looks at the world
in new ways . . .</td><td>Uses two systems to
organize . . .</td><td>Has two main branches
and other branches . . .</td></tr>
<tr><td>Definition
____
____
_______________________</td><td>Five Themes
—location, place, regions,
human environment,
interaction, movement</td><td>Physical
—study of world's
physical features</td></tr>
<tr><td>Tools
—Maps, globe, other
tools</td><td>Six Elements
—spatial, places and
regions, physical, human,
environment and society,
uses of geography</td><td>Human
—study of world's
people and cultures</td></tr>
<tr><td>Levels
—local, regional, global</td><td></td><td>Other Branches
—cartography,
hydrology, meteorology,
others</td></tr>
</table>

COMPREHENSION AND CRITICAL THINKING

Use information from the graphic organizer to answer the following
questions.

1. Identify Read the three definitions below. Write the one that best defines
geography on the lines under the first column. Geography is the study of . . .

    regions and their physical features.

    the world, people, and their landscapes.

    how different human cultures developed.

2. Analyze Give an example of how the five themes and six elements are similar.

____

____

3. Interpret If a geographer were making a map to show the size and height of a
hilly area, what branches of geography would this involve?

____

4. Draw a Conclusion How does geography help us to better understand the world
and its peoples?

____

A Geographer's World

Section 1

MAIN IDEAS

1. Geography is the study of the world, its people, and the landscapes they create.
2. Geographers look at the world in many different ways.
3. Maps and other tools help geographers study the planet.

Key Terms and Places

geography the study of the world, its people, and the landscapes they create

landscape the human and physical features that make a place unique

social science a field that studies people and the relationships among them

region a part of the world with one or more common features distinguishing it from surrounding areas

map a flat drawing that shows part of Earth's surface

globe a spherical model of the entire planet

Section Summary

WHAT IS GEOGRAPHY?

For every place on Earth, you can ask questions to learn about it: What does the land look like? What is the weather like? What are people's lives like? Asking questions like these is how you study geography. **Geography** is the study of the world, its people, and the **landscapes** they create.

Geographers (people who study geography) ask questions about how the world works. For example, they may ask why a place gets tornadoes. To find answers, they gather data by observing and measuring. In this way, geography is like science.

> Underline the sentence that states how geography is like science.

Geography can also be like a social science. **Social science** studies people and how they relate to each other. This information cannot be measured in the same way. To study people, geographers may visit places and talk to the people about their lives.

LOOKING AT THE WORLD

Geographers must look carefully at the world around them. Depending on what they want to learn, they look at the world at different levels.

Geographers may study at the local level, such as a city or town. They may ask why people live there, what work they do, and how they travel. They can help a town or city plan improvements.

Geographers may also study at the regional level. A **region** is an area with common features. A region may be big or small. Its features make it different from areas around it. The features may be physical (such as mountains) or human (such as language).

Sometimes geographers study at the global level. They study how people interact all over the world. Geographers can help us learn how people's actions affect other people and places. For example, they may ask how one region influences other regions.

> Circle the three levels that geographers study.

THE GEOGRAPHER'S TOOLS

Geographers need tools to do their work. Often, they use maps and globes. A **map** is a flat drawing that shows Earth's surface. A **globe** is a spherical (round) model of the whole planet.

Maps and globes both show what Earth looks like. Because a globe is round, it can show Earth as it really is. To show the round Earth on a flat map, some details have to change. For example, a place's shape may change a little. But maps have benefits. They are easier to work with. They can also show small areas, such as cities, better.

Geographers also use other tools, such as satellite images, computers, notebooks, and tape recorders.

> In what way are maps and globes similar?
>
> _______________________________
>
> _______________________________

> Underline two sentences that tell the benefits of using maps.

CHALLENGE ACTIVITY

Critical Thinking: Solving Problems Pick a foreign country you would like to study. You want to develop the most complete picture possible of this place and its people. Make a list of questions to ask and tools you would use to find the answers.

                          Interactive Reader and Study Guide

A Geographer's World

Section 2

MAIN IDEAS

1. The five themes of geography help us organize our studies of the world.

2. The six essential elements of geography highlight some of the subject's most important ideas.

Key Terms and Places

absolute location a specific description of where a place is

relative location a general description of where a place is

environment an area's land, water, climate, plants and animals, and other physical features

Section Summary

THE FIVE THEMES OF GEOGRAPHY

Geographers use themes in their work. A theme is a topic that is common throughout a discussion or event. Many holidays have a theme, such as the flag and patriotism on the Fourth of July.

There are five major themes of geography: Location, Place, Human-Environment Interaction, Movement, and Regions. Geographers can use these themes in almost everything they study.

Location describes where a place is. This may be specific, such as an address. This is called an **absolute location**. It may also be general, such as saying the United States is north of Central America. This is called a **relative location**.

Place refers to an area's landscape. The landscape is made up of the physical and human features of a place. Together, these features give a place its own identity apart from other places.

Human-Environment Interaction studies how people and their environment affect each other. The **environment** includes an area's physical features, such as land, water, weather, and animals.

> List the five major themes of geography:
>
> _______________________
>
> _______________________
>
> _______________________
>
> _______________________
>
> _______________________

Geographers study how people change their environment (by building, for example). They also study how the environment causes people to adapt (by dressing for the weather, for example).

Movement involves learning about why and how people move. Do they move for work or pleasure? Do they travel by roads or other routes?

Studying Regions helps geographers learn how places are alike and different. This also helps them learn why places developed the way they did.

> **Describe two ways that people and their environment affect each other.**
>
> ___________________
>
> ___________________
>
> ___________________
>
> ___________________
>
> ___________________

THE SIX ESSENTIAL ELEMENTS

It is important to organize how you study geography, so you get the most complete picture of a place. Using the five major themes can help you do this. Using the six essential elements can, also.

> **What do the five themes and six elements of geography help you do? Underline the sentence that explains this.**

Geographers and teachers created the six elements from eighteen basic ideas, called standards. The standards say what everyone should understand about geography. Each element groups together the standards that are related to each other.

The six elements are: The World in Spatial Terms (spatial refers to where places are located); Places and Regions; Physical Systems; Human Systems; Environment and Society; Uses of Geography. The six elements build on the five themes, so some elements and themes are similar. Uses of Geography is not part of the five themes. It focuses on how people can use geography to learn about the past and present, and plan for the future.

CHALLENGE ACTIVITY

Critical Thinking: Analyze Analyze a place you regularly visit, such as a vacation spot or a park in your neighborhood. Write a question about the place for each geography theme to help someone not familiar with the themes understand them.

 Interactive Reader and Study Guide

A Geographer's World

Section 3

MAIN IDEAS

1. Physical geography is the study of landforms, water bodies, and other physical features.

2. Human geography focuses on people, their cultures, and the landscapes they create.

3. Other branches of geography examine specific aspects of the physical or human world.

Key Terms and Places

physical geography the study of the world's physical features, such as landforms, bodies of water, climates, soils, and plants

human geography the study of the world's people, communities, and landscapes

cartography the science of making maps

meteorology the study of weather and what causes it

Section Summary

PHYSICAL GEOGRAPHY

The field of geography has many branches, or divisions. Each branch has a certain focus. No branch alone gives us a picture of the whole world. When looked at together, the different branches help us understand Earth and its people better.

Geography has two main branches: physical geography and human geography. **Physical geography** is the study of the world's physical features, such as landforms, bodies of water, and weather.

Physical geographers ask questions about Earth's many physical features: Where are the mountains and flat areas? Why are some areas rainy and others dry? Why do rivers flow a certain way? To get their answers, physical geographers measure features— such as heights of mountains and temperatures of places.

> What do the different branches of geography help us do when they are looked at together? Underline the sentence that answers this.

> List the two main branches of geography:
>
> _______________________
>
> _______________________

                                      Interactive Reader and Study Guide

Physical geography has important uses. It helps us understand how the world works. It also helps us predict and prepare for dangers, such as big storms.

HUMAN GEOGRAPHY

Human geography is the study of the world's people, communities, and landscapes. It is the other main branch of geography.

Human geographers study people in the past or present. They ask why more people live in some places than in others. They also ask other questions, such as what kinds of work people do.

People all over the world are very different, so human geographers often study a smaller topic. They might study people in one region, such as central Africa. They might study one part of people's lives in different regions, such as city life.

Human geography also has important uses. It helps us learn how people meet basic needs for food, water, and shelter. It helps people improve their lives. It can also help protect the environment.

> Why do human geographers often study one smaller topic?

> Circle three basic needs that people have to meet.

OTHER FIELDS OF GEOGRAPHY

Other branches of geography study one aspect of the world. Some of these are smaller parts of physical geography or of human geography.

Here are a few other branches to know about. **Cartography** is the science of making maps. Hydrology is the study of water on Earth. **Meteorology** is the study of weather and what causes it.

> What is meteorology?

CHALLENGE ACTIVITY

Critical Thinking: Drawing Inferences Examine a map of an unfamiliar city using a road atlas or an online map. Write a paragraph telling a visitor what physical and human features to look for in each quadrant (NE, SE, NW, SW).

Interactive Reader and Study Guide

Planet Earth

CHAPTER SUMMARY

CAUSE		EFFECT
Earth's rotation	→	day and night
Earth's revolution	→	the seasons
the water cycle	→	a constant supply of water on Earth
________________	→	water shortages
continental drift	→	creates landforms
wind, water, and ice	→	________________

COMPREHENSION AND CRITICAL THINKING

Use information from the graphic organizer to answer the following questions.

1. Recall What is the effect of Earth's movement in space?

____

____

2. Interpret Which two of the following words belong in the graphic organizer above: drought, tilt, precipitation, erosion? Write the two words in the spaces provided.

3. Synthesis Under what general theme do all of these cause-and-effect relationships fall?

____

____

Planet Earth

Section 1

MAIN IDEAS

1. Earth's movement affects the amount of energy we receive from the sun.
2. Earth's seasons are caused by the planet's tilt.

Key Terms

solar energy energy from the sun

rotation one complete spin of Earth on its axis

revolution one trip of Earth around the sun

latitude the distance north or south of Earth's equator

tropics regions close to the equator

Section Summary

EARTH'S MOVEMENT

Energy from the sun, or **solar energy**, is necessary for life on Earth. It helps plants grow and provides light and heat. Several factors affect the amount of solar energy Earth receives. These are rotation, revolution, tilt, and latitude.

Earth's axis is an imaginary rod running from the North Pole to the South Pole. Earth spins around on its axis. One complete **rotation** takes 24 hours, or one day. Solar energy reaches only half of the planet at a time. As Earth rotates, levels of solar energy change. The half that faces the sun receives light and heat and is warmer. The half that faces away from the sun is darker and cooler.

As Earth rotates, it also moves around the sun. Earth completes one **revolution** around the sun every year, in 365 1/4 days. Every four years an extra day is added to February. This makes up for the extra quarter of a day.

Earth's axis is tilted, not straight up and down. At different times of year, some locations tilt toward the sun. They get more solar energy than locations tilted away from the sun.

> Name the four factors that affect the amount of solar energy Earth receives.
>
> ________________________
>
> ________________________

> What would happen if Earth did not rotate?
>
> ________________________
>
> ________________________

> Underline the sentence that describes Earth's revolution around the sun.

> Where you live, does more solar energy reach Earth in winter or in summer?
>
> ________________________

 Interactive Reader and Study Guide

Section 1, continued

Latitude refers to imaginary lines that run east and west around the planet, north and south of the Earth's equator. Areas near the equator receive direct rays from the sun all year and have warm temperatures. Higher latitudes receive fewer direct rays and are cooler.

> Why are areas near the equator warmer than those in higher latitudes?

THE SEASONS

Many locations on Earth have four seasons: winter, spring, summer, and fall. These are based on temperature and how long the days are.

The seasons change because of the tilt of Earth's axis. In summer the Northern Hemisphere is tilted toward the sun. It receives more solar energy than during the winter, when it is tilted away from the sun.

Because Earth's axis is tilted, the hemispheres have opposite seasons. Winter in the Northern Hemisphere is summer in the Southern Hemisphere. During the fall and spring, the poles point neither toward nor away from the sun. In spring, temperatures rise and days become longer as summer approaches. In fall the opposite occurs.

> What would the seasons be like in the Northern and Southern hemispheres if Earth's axis weren't tilted?

In some regions, the seasons are tied to rainfall instead of temperature. One of these regions, close to the equator, is the **tropics**. There, winds bring heavy rains from June to October. The weather turns dry in the tropics from November to January.

> Circle the name of the warm region near the equator.

CHALLENGE ACTIVITY

Critical Thinking: Drawing Conclusions Imagine that you are a travel agent. One of your clients is planning a trip to Argentina in June, and another is planning a trip to Chicago in August. What kinds of clothing would you suggest they pack for their trips and why?

Planet Earth

Section 2

MAIN IDEAS

1. Salt water and freshwater make up Earth's water supply.
2. In the water cycle, water circulates from Earth's surface to the atmosphere and back again.
3. Water plays an important role in people's lives.

Key Terms

freshwater water without salt

glacier large area of slow-moving ice

surface water water that is stored in Earth's streams, rivers, and lakes

precipitation water that falls to Earth's surface as rain, snow, sleet, or hail

groundwater water found below Earth's surface

water vapor water that occurs in the air as an invisible gas

water cycle the circulation of water from Earth's surface to the atmosphere and back

drought a long period of lower-than-normal precipitation

Section Summary

EARTH'S WATER SUPPLY

Approximately three-quarters of Earth's surface is covered with water. There are two kinds of water—salt water and **freshwater**. About 97 percent of Earth's water is salt water. Most of it is in the oceans, seas, gulfs, bays, and straits. Some lakes, such as the Great Salt Lake in Utah, also contain salt water.

> Circle the places where we find salt water.

Salt water cannot be used for drinking. Only freshwater is safe to drink. Freshwater is found in lakes and rivers and stored underground. Much is frozen in **glaciers**. Freshwater is also found in the ice of the Arctic and Antarctic regions.

> Use an atlas or a globe to locate the Great Salt Lake, the Arctic, and the Antarctic regions.

One form of freshwater is **surface water**. This is stored in streams, lakes, and rivers. Streams form when **precipitation** falls to Earth as rain, snow, sleet, or hail. These streams then flow into larger streams and rivers.

Most freshwater is stored underground.
Groundwater bubbles to the surface in springs or
can be reached by digging deep holes, or wells.

THE WATER CYCLE

Water can take the form of a liquid, gas, or solid. In
its solid form, water is snow and ice. Liquid water is
rain or water found in lakes and rivers. **Water vapor**
is an invisible form of water in the air.

Water is always moving. When water on Earth's
surface heats up, it evaporates and turns into water
vapor. It then rises from Earth into the atmosphere.
When it cools down, it changes from water vapor to
liquid. Droplets of water form clouds. When they get
heavier, these droplets fall to Earth as precipitation.
This process of evaporation and precipitation is
called the **water cycle**.

Some precipitation is absorbed into the soil as
groundwater. The rest flows into streams, rivers,
and oceans.

WATER AND PEOPLE

Problems with water include shortages, pollution,
and flooding. Shortages are caused by overuse and
by **drought**, when there is little or no precipitation
for a long time. Chemicals and waste can pollute
water. Heavy rains can cause flooding.

Water quenches our thirst and allows us to have
food to eat. It is an important source of energy.
Water also provides recreation, making our lives
richer and more enjoyable. Water is essential for life
on Earth.

CHALLENGE ACTIVITY

Critical Thinking: Solving Problems You are campaigning for public
office. Write a speech describing three actions you plan to take to
protect supplies of freshwater.

> Underline the words that define water vapor.

> What are the two main processes of the water cycle?

> What water problems affect human beings?

Planet Earth

Section 3

MAIN IDEAS

1. Earth's surface is covered by many different landforms.
2. Forces below Earth's surface build up our landforms.
3. Forces on the planet's surface shape Earth's landforms.
4. Landforms influence people's lives and culture.

Key Terms

landforms shapes on Earth's surface, such as hills or mountains

continents large landmasses

plate tectonics a theory suggesting that Earth's surface is divided into more than 12 slow-moving plates, or pieces of Earth's crust

lava magma, or liquid rock, that reaches Earth's surface

earthquake sudden, violent movement of Earth's crust

weathering the process of breaking rock into smaller pieces

erosion the movement of sediment from one location to another

Section Summary

LANDFORMS

Geographers study **landforms** such as mountains, valleys, plains, islands, and peninsulas. They study how landforms are made and how they influence people.

> Give two examples of landforms.
> _______________________
> _______________________

FORCES BELOW EARTH'S SURFACE

Below Earth's surface, or crust, is a layer of liquid and a solid core. The planet has seven **continents**, large landmasses made of Earth's crust. All of Earth's crust rests on 12 plates. These plates are constantly in motion. Geographers call the study of these moving pieces of crust **plate tectonics**.

All of these plates move at different speeds and in different directions. As they move, they shape Earth's landforms. Plates move in three ways: They collide, they separate, and they slide past each other.

The energy of colliding plates creates new landforms. When two ocean plates collide, they may

> Look in an atlas. How might the theory of plate tectonics explain the shapes of North America, South America, and Africa?
> _______________________

    Interactive Reader and Study Guide

Section 3, *continued*

form deep valleys on the ocean's floor. When ocean plates collide with continental plates, mountain ranges are formed. Mountains are also created when two continental plates collide.

When plates separate, usually on the ocean floor, they cause gaps in the planet's crust. Magma, or liquid rock, rises through the cracks as **lava.** As it cools, it forms underwater mountains or ridges. Sometimes these mountains rise above the surface of the water and form islands.

Plates can also slide past each other. They grind along faults, causing **earthquakes.**

> Underline what happens when ocean plates collide with one another.

> What causes earthquakes?
> _______________________
> _______________________

FORCES ON EARTH'S SURFACE

As landforms are created, other forces work to wear them away. **Weathering** breaks larger rocks into smaller rocks. Changes in temperature can cause cracks in rocks. Water then gets into the cracks, expands as it freezes, and breaks the rocks. Rocks eventually break down into smaller pieces called sediment. Flowing water moves sediment to form new landforms, such as river deltas.

Another force that wears down landforms is **erosion.** Erosion takes place when sediment is moved by ice, water, and wind.

> Circle the three forces that can cause erosion.

LANDFORMS INFLUENCE LIFE

Landforms influence where people live. For example, people might want to settle in an area with good soil and water. People change landforms in many ways. For example, engineers build tunnels through mountains to make roads. Farmers build terraces on steep hillsides.

CHALLENGE ACTIVITY

Critical Thinking: Drawing Inferences Find out about a landform in your area that was changed by people. Write a report explaining why and how it was changed.

                                    Interactive Reader and Study Guide

Climate, Environment, and Resources

CHAPTER SUMMARY

CLIMATE	ENVIRONMENT	RESOURCES
front	deforestation	petroleum
rain shadow	savanna	solar energy
ocean currents	habitat	renewable resources
precipitation	desert	minerals
prevailing winds	ecosystem	fossil fuels
dry air	humus	soil
high pressure	rain forest	forest
hurricane	pollution	erosion
sunlight	extinct	hydroelectric power
Gulf Stream	river	water

COMPREHENSION AND CRITICAL THINKING

Use information from the graphic organizer to answer the following questions.

1. Explain Choose a term from the *Environment* column that is affected by hydroelectric power. Write a sentence explaining how it is affected.

____

____

2. Interpret Under which other column could the word *desert* be listed? Why?

____

____

3. Identify Cause and Effect Explain how desertification occurs and why it is a problem.

____

____

____

____

Climate, Environment, and Resources

Section 1

MAIN IDEAS

1. While weather is short term, climate is a region's average weather over a long period.
2. The amount of sun at a given location is affected by Earth's tilt, movement, and shape.
3. Wind and water move heat around Earth, affecting how warm or wet a place is.
4. Mountains influence temperature and precipitation.

Key Terms and Places

weather the short-term changes in the air for a given place and time

climate a region's average weather conditions over a long period

prevailing winds winds that blow in the same direction over large areas of Earth

ocean currents large streams of surface seawater

front a place where two air masses of different temperature or moisture content meet

Section Summary

UNDERSTANDING WEATHER AND CLIMATE

Weather is the condition of the atmosphere at a certain time and place. **Climate** is a region's average weather over a long time. Climate is affected mostly by two factors: sun and latitude. Energy from the sun falls more directly on the equator, so the hottest temperatures are near the equator. In general it gets colder as you move away from the equator to a higher latitude.

> **What are two important forces that affect climate?**
>
> _______________________________
>
> _______________________________

SUN AND LOCATION

Heat from the sun moves around the Earth, partly through winds. Wind is caused by the rising and sinking of air. Cold air sinks, and warm air rises. At different latitudes winds tend to blow in the same direction. These **prevailing winds** can be from the west or east. Near the poles and in the subtropics, prevailing winds are easterlies. In the middle latitudes are the westerlies. Prevailing winds control an area's climate.

> **What causes wind?**
>
> _______________________________
>
> _______________________________

  Interactive Reader and Study Guide

Section 1, continued

WIND AND WATER

Winds pick up moisture over oceans and dry out passing over land. At about 30° North and South latitude, dry winds cause many of the world's deserts.

Ocean currents—large streams of surface water—also move heat around. The Gulf Stream is a warm current that flows from the Gulf of Mexico across the Atlantic Ocean to western Europe.

Water heats and cools more slowly than land. Therefore, water helps to moderate the temperature of nearby land, keeping it from getting very hot or very cold.

A **front** is a place where two different air masses meet. In the United States and other regions, warm and cold air masses meet often, causing severe weather. These can include thunderstorms, blizzards, and tornadoes. Tornadoes are twisting funnels of air that touch the ground. Hurricanes are large tropical storms that form over water. They bring destructive high winds and heavy rain.

> **Which heats and cools more slowly—land or water?**
> ___________________

> **Which occurs in the tropics— tornadoes or hurricanes?**
> ___________________

MOUNTAINS

Mountains also affect climate. Warm air blowing against a mountainside rises and cools. Clouds form, and precipitation falls on the side facing the wind. However, the air is dry by the time it goes over the mountain. This effect creates a rain shadow, a dry area on the side of the mountain facing land.

CHALLENGE ACTIVITY

Critical Thinking: Sequencing Write a short description of the process leading up to the formation of a rain shadow. Draw and label a picture to go with your description.

Climate, Environment, and Resources

Section 2

MAIN IDEAS

1. Geographers use temperature, precipitation, and plant life to identify climate zones.
2. Tropical climates are wet and warm, while dry climates receive little or no rain.
3. Temperate climates have the most seasonal change.
4. Polar climates are cold and dry, while highland climates change with elevation.

Key Terms and Places

monsoon winds that shift direction with the seasons and create wet and dry periods

savanna an area of tall grasses and scattered trees and shrubs

steppe a semi-dry grassland or prairie

permafrost permanently frozen layers of soil

Section Summary

MAJOR CLIMATE ZONES

We can divide Earth into five climate zones: tropical, temperate, polar, dry, and highland. Tropical climates appear near the equator, temperate climates are found in the middle latitudes, and polar climates occur near the poles. Dry and highland climates can appear at different latitudes.

> Underline the names of the five climate zones.

TROPICAL AND DRY CLIMATES

Humid tropical climates occur near the equator. Some are warm and rainy throughout the year. Others have **monsoons**—winds that shift directions and create wet and dry seasons. Rain forests need a humid climate to thrive and support thousands of species.

> What happens when monsoon winds change direction?
>
> _______________________________
> _______________________________

Moving away from the equator, we find tropical savanna climates. A **savanna** is an area of tall grasses and scattered trees and shrubs. A long, hot dry season is followed by short periods of rain.

Deserts are hot and dry. At night, the dry air cools quickly; desert nights can be cold. Only a few living things survive in a desert. Sometimes **steppes**—dry grasslands—are found near deserts.

TEMPERATE CLIMATES

Away from the ocean in the middle latitudes are humid continental climates. Most have four distinct seasons, with hot summers and cold winters. In this climate, weather often changes quickly when cold and warm air masses meet.

A Mediterranean climate has hot, sunny summers and mild, wet winters. They occur near the ocean, and the climate is mostly pleasant. People like to vacation in these climates. Only small, scattered trees survive in these areas.

East coasts near the tropics have humid subtropical climates, because of winds bringing in moisture from the ocean. They have hot, wet summers and mild winters. Marine west coast climates occur farther north and also get moisture from prevailing winds coming in from the ocean.

> **Underline the name of the climate that can have four distinct seasons.**

> **What do people typically like to do in Mediterranean climates?**
> _______________________________

POLAR AND HIGHLAND CLIMATES

Subarctic climate occurs south of the Arctic Ocean. Winters are long and cold, and summers are cool. There is enough precipitation to support forests. At the same latitude near the coasts, tundra climate is also cold, but too dry for trees to survive. In parts of the tundra, soil is frozen as **permafrost**.

Ice cap climates are the coldest on Earth. There is little precipitation and little vegetation.

Highland, or mountain, climate changes with elevation. As you go up a mountain, the climate may go from tropical to polar.

> **Can there be forests in subarctic climates?**
> _______________________________

CHALLENGE ACTIVITY

Critical Thinking: Comparing and Contrasting Create a table showing the differences and similarities between any two types of climate.

Climate, Environment, and Resources

Section 3

MAIN IDEAS

1. The environment and life are interconnected and exist in a fragile balance.
2. Soils play an important role in the environment.

Key Terms and Places

environment a plant or animal's surroundings

ecosystem any place where plants and animals depend upon each other and their environment for survival

habitat the place where a plant or animal lives

extinct to die out completely

humus decayed plant or animal matter

desertification the slow process of losing soil fertility and plant life

Section Summary

THE ENVIRONMENT AND LIFE

Plants and animals cannot live just anywhere. They must have the right surroundings, or **environment**. Climate, land features, and water are all part of a living thing's environment. If an area has everything a living thing needs, it can be a **habitat** for that species.

Many plants and animals usually share a habitat. Small animals eat plants, and then large animals eat the small animals. Species are connected in many ways. A community of connected species is called an **ecosystem**. Ecosystems can be as small as a pond or as large as the entire Earth.

Geographers study how changes in environments affect living things. Natural events and human actions change environments. Natural events include forest fires, disease, and climate changes. Human actions include clearing land and polluting.

> **How large can an ecosystem be?**
> ________________________

> **Underline the human actions that can cause changes in an environment.**

Section 3, *continued*

If a change to the environment is extreme, a species might become **extinct**, or die out completely.

> What does it mean to become extinct?

SOILS AND THE ENVIRONMENT

Without soil, much of our food would not exist. Soil forms in layers over hundreds or thousands of years. The most fertile layer, the topsoil, has the most humus. **Humus** is decayed plant or animal matter.

> Which has more humus—topsoil or subsoil?

The next layer, the subsoil, has less humus and more material from rocks. Soil gets minerals from these rocks. Below the subsoil is mostly rock.

An environment's soil affects which plants can grow there. Fertile soils have lots of humus and minerals. Fertile soils also need to contain water and small air spaces.

> Underline two things found in fertile soil.

Soils can lose fertility from erosion by wind or water. Soil can also lose fertility from planting the same crops repeatedly. If soil becomes worn out and can no longer support plants, **desertification** can occur.

CHALLENGE ACTIVITY

Critical Thinking: Drawing Inferences Consider the interconnections in your environment. As you go through a normal day, keep a list of the sources you rely on for energy, food, and water.

Climate, Environment, and Resources

Section 4

MAIN IDEAS

1. Earth provides valuable resources for our use.
2. Energy resources provide fuel, heat, and electricity.
3. Mineral resources include metals, rocks, and salt.
4. Resources shape people's lives and countries' wealth.

Key Terms and Places

natural resource any material in nature that people use and value

renewable resources resources that can be replaced naturally

nonrenewable resources resources that cannot be replaced

deforestation the loss of forestland

reforestation planting trees to replace lost forestland

fossil fuels nonrenewable resources formed from the remains of ancient plants and animals

hydroelectric power the production of electricity by moving water

Section Summary

EARTH'S VALUABLE RESOURCES

Anything in nature that people use and value is a **natural resource**. These include such ordinary things as air, water, and soil. Resources such as trees are called **renewable resources** because Earth replaces them naturally. Those that cannot be replaced, such as oil, are called **nonrenewable resources**.

Air and water are renewable resources, but pollution can damage both. Some people get their water from underground wells, which can run out if too many people use them.

Soil is needed for all plant growth, including trees in forests. We get lumber, medicine, nuts, and rubber from forests. Soil and trees are renewable, but must be protected. The loss of forests is called **deforestation**. When we plant trees to replace lost forests, we call it **reforestation**.

> Are air and water renewable or nonrenewable resources?
>
> _______________________

> Underline the resources we can get from a forest.

Section 4, *continued*

ENERGY RESOURCES

Most of our energy comes from **fossil fuels**, which
are formed from the remains of ancient living things.
These include coal, oil, and natural gas.

We use coal mostly for electricity, but it causes
air pollution. An advantage of coal is that Earth still
has a large supply. Another fossil fuel is petroleum,
or oil. It is used to make gasoline and heating oil.
Oil can be turned into plastics and other products.
Oil also causes pollution, but we depend on it for
much of our energy. The cleanest fossil fuel is natural
gas, which is used mainly for cooking and heating.

Renewable energy resources include
hydroelectric power—the creation of electricity
by moving water. This is accomplished mainly by
building dams on rivers. Other renewable energy
sources are wind, solar, and nuclear energy. Nuclear
energy produces dangerous waste material that
must be stored for thousands of years.

MINERAL RESOURCES

Minerals are solid substances in the Earth's crust
formed from nonliving matter. Like fossil fuels,
minerals are nonrenewable. Types of minerals include
metals, rocks and gemstones, and salt. Mineral uses
include making steel from iron, making window
glass from quartz, and using stone as a building
material. We also use minerals to make jewelry,
coins, and many other common objects.

RESOURCES AND PEOPLE

Some places are rich in natural resources. Resources
such as fertile farmland, forests, and oil have helped
the United States become a powerful country with a
strong economy.

CHALLENGE ACTIVITY

Critical Thinking: Drawing Inferences Write a short essay explaining
why we still use coal, even though it causes pollution.

          Interactive Reader and Study Guide

The World's People

CHAPTER SUMMARY

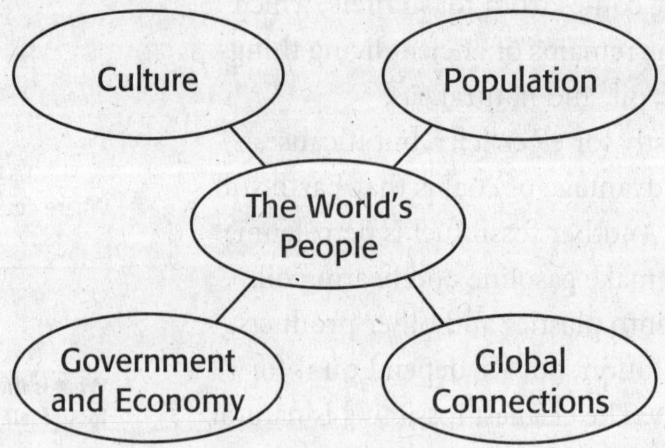

COMPREHENSION AND CRITICAL THINKING

Use information from the graphic organizer to answer the following questions.

1. Identify What four types of government would you list in the "Government and Economy" circle to make the graphic organizer more detailed?

____

2. Analyze Give an example of how the population in an area can affect the economy.

____

____

3. Make an Inference How can global connections affect the culture in an area?

____

____

4. Draw a Conclusion If the world's population continues to grow rapidly, what might be the most important challenges for people to overcome?

____

____

____

The World's People

Section 1

MAIN IDEAS

1. Culture is the set of beliefs, goals, and practices that a group of people share.
2. The world includes many different culture groups.
3. New ideas and events lead to changes in culture.

Key Terms

culture the set of beliefs, values, and practices a group of people have in common
culture trait an activity or behavior in which people often take part
culture region an area in which people have many shared culture traits
ethnic group a group of people who share a common culture and ancestry
cultural diversity having a variety of cultures in the same area
cultural diffusion the spread of culture traits from one region to another

Section Summary

WHAT IS CULTURE?

Culture is the set of beliefs, values, and practices a group of people have in common. Everything in day-to-day life is part of culture, including language, religion, clothes, music, and foods. People everywhere share certain basic cultural features, such as forming a government, educating children, and creating art or music. However, people practice these things in different ways, making each culture unique.

Culture traits are activities or behaviors in which people often take part, such as language and popular sports. People share some culture traits, but not others. For example, people eat using forks, chopsticks, or their fingers in different areas.

CULTURE GROUPS

There are thousands of different cultures in the world. People who share a culture are part of a culture group that may be based on things like age or religion.

Section 1, *continued*

A **culture region** is an area in which people have many shared culture traits such as language, religion, or lifestyle. One large culture region is the Arab world in Southwest Asia and North Africa. A country may have several different culture regions, or just a single region, such as Japan.

A culture region may be based on an **ethnic group**, a group of people who share the same religion, traditions, language, or foods. **Cultural diversity** is having a variety of cultures in the same area. It can create a variety of ideas and practices, but it can also lead to conflict.

> How can cultural diversity affect the people in an area?
>
> _______________________________
>
> _______________________________

CHANGES IN CULTURE

Cultures are constantly changing. They can change through the development of new ideas or contact with other societies. New ideas such as the development of electricity, motion pictures, and the Internet have changed what people do and how they communicate. When two cultures come in close contact, both usually change. For example, the Spanish and Native American cultures changed when the Spanish conquered the Americas.

> Underline the sentences that describe how cultures change.

Cultural diffusion is the spread of culture traits from one part of the world to another. It can happen when people move and bring their culture with them. New ideas and customs, such as baseball or clothing styles, can spread from one place to another as people learn about them.

> What are two ways cultural diffusion occurs?
>
> _______________________________

CHALLENGE ACTIVITY

Critical Thinking: Drawing Inferences Consider all of the parts of your culture that have been influenced by other cultures. During a normal day, keep a list of all the things you use or do that you think have been influenced by other cultures.

The World's People

Section 2

MAIN IDEAS

1. The study of population patterns helps geographers learn about the world.
2. Population statistics and trends are important measures of population change.

Key Terms

population the total number of people in a given area

population density a measure of the number of people living in an area, usually expressed as persons per square mile or square kilometer

birthrate the annual number of births per 1,000 people

migration the process of moving from one place to live in another

Section Summary

POPULATION PATTERNS

Population is the total number of people in a given area. Geographers study population patterns to learn about the world.

Some places are crowded with people, while others are almost empty. **Population density** is a measure of the number of people living in an area, usually expressed as persons per square mile or square kilometer. It describes how crowded a place is, which in turn affects how people live. In places with a high density, there is little open space, buildings are taller, and roads are more crowded than places with lower density. They also often have more products available for a variety of shoppers.

High density areas often have fertile soil, available water, and a favorable climate for agriculture. Areas that are less dense often have harsh land or climate that makes survival harder.

> Underline the two sentences that describe the effects of population density on a place.

> What is the land and climate often like in areas of high population density?
>
> _______________________________
> _______________________________

POPULATION CHANGE

The number of people living in an area affects jobs, housing, schools, medical care, available food, and many other things. Geographers study population

  Interactive Reader and Study Guide

Section 2, *continued*

changes and world trends to understand how people live.

Three statistics are important to study a country's population over time. **Birthrate** is the annual number of births per 1,000 people. Death rate is the annual number of deaths per 1,000 people. The rate of natural increase is found by subtracting the death rate from the birthrate.

> Underline the sentence that tells how to calculate the rate of natural increase.

Some areas have low rates of natural increase, such as Europe and North America. Some countries in Africa and Asia have very high rates of natural increase. High rates make it hard for countries to develop economically because they need to provide jobs, education, and medical care for a growing population.

> Why do high rates of natural increase make it hard for a country to develop economically?
>
> _______________________________
>
> _______________________________

Migration is the process of moving from one place to live in another. People may leave a place because of problems there, such as war, famine, drought, or lack of jobs. Other people may move to find political or religious freedom or economic opportunities in a new place.

The world's population has grown very rapidly in the last 200 years. Better health care and food supplies have helped more babies survive and eventually have children of their own. Many industrialized countries currently have slow population growth while other countries have very fast growth. Fast growth puts a strain on resources, housing, and government aid.

> How has the world's population changed during the last 200 years?
>
> _______________________________
>
> _______________________________

CHALLENGE ACTIVITY

Critical Thinking: Identify Cause and Effect Find out the population density of your city or town. Write down ways that this density affects your life and the lives of others.

The World's People

Section 3

MAIN IDEAS

1. The governments of the world include democracy, monarchy, dictatorship, and communism.
2. Different economic activities and systems exist throughout the world.
3. Geographers group the countries of the world based on their level of economic development.

Key Terms

democracy a form of government in which the people elect leaders and rule by majority

communism a political system in which the government owns all property and dominates all aspects of life in the country

market economy a system based on private ownership, free trade, and competition

command economy a system in which the central government makes all economic decisions

gross domestic product (GDP) the value of all goods and services produced within a country in a single year

developed countries countries with strong economies and a high quality of life

developing countries countries with less productive economies and a lower quality of life

Section Summary

GOVERNMENTS OF THE WORLD

People form governments to make laws, regulate business, and provide aid to people. A **democracy** is a form of government in which the people elect leaders and rule by majority. Most democracies protect people's rights to freedom of speech, religion, and a free press.

> **What rights are protected in most democracies?**
> _______________________
> _______________________

Monarchies are ruled directly by a king or queen who holds all the power. Dictatorships are also ruled by a single person. Dictators hold all the power and often rule by force. **Communism** is a system in which the government owns all property and dominates all aspects of life in the country. In most Communist states, the people have restricted rights and little freedom.

> **What types of government are ruled by a single person?**
> _______________________
> _______________________

ECONOMIES OF THE WORLD

Primary industries provide natural resources to others through work such as farming, fishing, and mining. Secondary industries use raw materials to manufacture other products such as automobiles or furniture. Tertiary industries exchange goods and services through retail stores, health care and educational organizations, and so on. Quaternary industries involve workers such as architects, lawyers, and scientists who research and distribute information.

In a traditional economy, people make and use their own goods with little exchange of goods. A **market economy** is based on free trade and competition. People buy and sell as they wish and prices are determined by supply and demand. In a **command economy**, the government decides what to produce and what prices will be.

> **What are the four levels of economic activity?**
> ________________________
> ________________________

> **Underline the sentence that describes who controls a command economy.**

ECONOMIC DEVELOPMENT

One measure of economic development is **gross domestic product** (**GDP**), the value of all goods and services produced within a country in a single year. Other ways include the level of industrialization and the quality of life.

Developed countries have strong, wealthy economies and high standards of living. **Developing countries** have poorer economies and a lower quality of life. About two thirds of the world's people live in developing countries with poor education and little access to health care or telecommunications.

> **Which type of country is more industrialized?**
> ________________________
> ________________________

CHALLENGE ACTIVITY

Critical Thinking: Classify Select a family member or friend and ask them about their job. Classify it as one of the four basic types and describe why it is important in our society.

         Interactive Reader and Study Guide

The World's People

Section 4

MAIN IDEAS

1. Globalization links the world's countries together through culture and trade.
2. The world community works together to solve global conflicts and crises.

Key Terms

globalization the process in which countries are increasingly linked to each other through culture and trade

popular culture culture traits that are well known and widely accepted

interdependence the reliance of one country on the resources, goods, or services of another country

United Nations (UN) an organization of the world's countries that promotes peace and security around the globe

humanitarian aid assistance to people in distress

Section Summary

GLOBALIZATION

People around the world are more closely linked than ever before. **Globalization** is the process in which countries are increasingly linked to each other through culture and trade. Improvements in technology and communication have increased globalization.

Popular culture consists of culture traits that are well known and widely accepted. These traits can include food, sports, music, and movies. The United States has a great influence on popular culture through sales of American products and the use of English for business, science, and education around the world. It is also greatly influenced by other countries.

World businesses are connected through trade. Companies may make products in many different countries or use products from around the world. **Interdependence** occurs when countries depend on each other for resources, goods, or services.

> Underline the sentence which describes two ways countries are linked together.

> What are four traits that can be considered part of popular culture?
>
> _______________________
>
> _______________________

   Interactive Reader and Study Guide

Companies and consumers depend on goods produced elsewhere.

A WORLD COMMUNITY

Because places around the world are connected closely, what happens in one place affects others. The world community works together to promote cooperation between countries.

When conflicts occur, countries from around the world try to settle them. The **United Nations (UN)** is an association of nearly 200 countries dedicated to promoting peace and security.

Crises such as earthquakes, floods, drought, or a tsunami can leave people in great need. Groups from around the world provide **humanitarian aid**, or assistance to people in distress. Some groups help refugees or provide medical care.

CHALLENGE ACTIVITY

Critical Thinking: Contrast Talk to a parent or other adult about their knowledge of other countries and their connections to them when they were young. Write a short essay that contrasts their global connections with yours.

History of the Fertile Crescent

CHAPTER SUMMARY

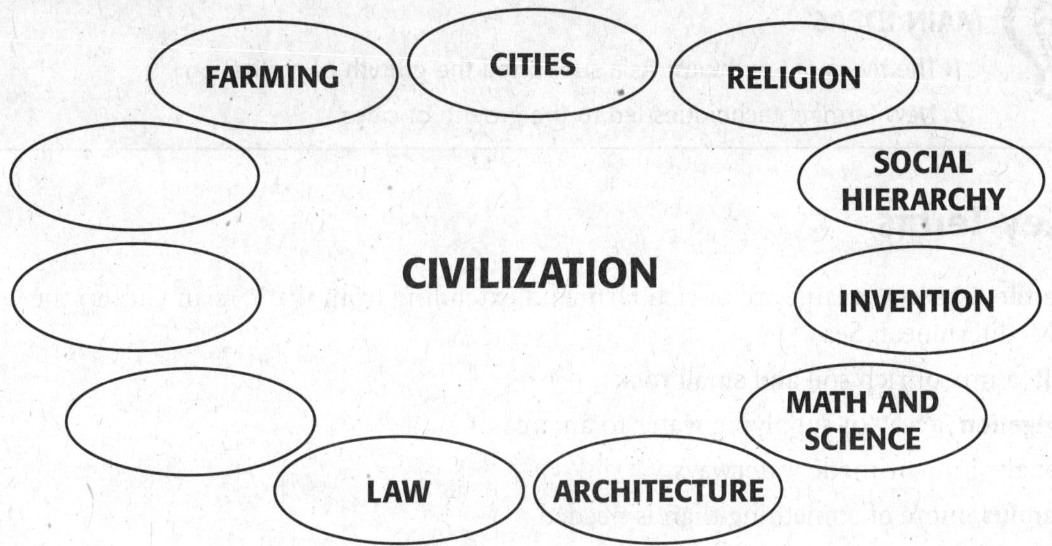

COMPREHENSION AND CRITICAL THINKING

Use information from the graphic organizer to answer the following questions.

1. Identify Which two of the eight aspects of civilization in the graphic organizer above can be classed as economic structures?

____

____

2. Interpret Of the following five terms (*the arts, silt, writing, trade, oil*), which three belong in the graphic organizer above? Write the three correct terms in the empty circles.

3. Sequence Of the eleven words or phrases around the word *civilization*, which comes first in time order? Which do you think comes last?

____

____

History of the Fertile Crescent

Section 1

MAIN IDEAS
1. The rivers of Southwest Asia supported the growth of civilization.
2. New farming techniques led to the growth of cities.

Key Terms

Fertile Crescent a large arc of rich farmland extending from the Persian Gulf to the Mediterranean Sea

silt a mix of rich soil and small rocks

irrigation a way of supplying water to an area of land

canals human-made waterways

surplus more of something than is needed

division of labor an arrangement in which people specialize in specific tasks

Section Summary

RIVERS SUPPORT THE GROWTH OF CIVILIZATION

Early people settled where crops would grow. Crops usually grew well near rivers, where water was available and regular floods made the soil rich.

Mesopotamia, part of the region known as the **Fertile Crescent** in Southwest Asia, lay between the Tigris and Euphrates rivers. Every year, floods on the rivers brought **silt**. The fertile silt made the land ideal for farming.

Hunter-gatherer groups first settled in Mesopotamia more than 12,000 years ago. Over time these people learned how to work together to control floods. They planted crops and grew their own food.

Farm settlements formed in Mesopotamia as early as 7000 BC. Farmers grew wheat, barley, and other grains. Livestock, birds, and fish were also sources of food. Plentiful food led to population growth and villages formed. Eventually, these early villages developed into the world's first civilization.

Mesopotamia means "between the rivers" in Greek. To which two rivers does the name of the region refer?

____

____

____

Name two grains grown by Mesopotamian farmers.

____

____

____

     Interactive Reader and Study Guide

Section 1, *continued*

FARMING AND CITIES

Early farmers faced the challenge of learning how to control the flow of river water to their fields in both rainy and dry seasons. Flooding destroyed crops, killed livestock, and washed away homes. When water levels were too low, crops dried up.

To solve their problems, Mesopotamians used **irrigation**. They dug out large storage basins to hold water supplies. Then they dug **canals** that connected these basins to a network of ditches. These ditches brought water to the fields and watered grazing areas for cattle and sheep.

Because irrigation made farmers more productive, they produced a **surplus**. Some people became free to do other jobs. For the first time, people became craftspersons, religious leaders, and government workers. A **division of labor** developed.

Mesopotamian settlements grew in size and complexity. Most people continued to work in farming jobs. However, cities became important places. People traded goods in cities. Cities became the political, religious, cultural, and economic centers of Mesopotamian civilization.

> **Underline the sentence that lists some of the problems caused by flooding.**

> **From where did the water collected in the storage basins come?**
>
> ________________________________
>
> ________________________________

> **Which places in Mesopotamia became the centers of civilization?**
>
> ________________________________
>
> ________________________________

CHALLENGE ACTIVITY

Critical Thinking: Drawing Inferences Write a proposal for an irrigation system that will divert flood waters and benefit riverbank farmers.

Interactive Reader and Study Guide

History of the Fertile Crescent

Section 2

MAIN IDEAS

1. The Sumerians created the world's first advanced society.
2. Religion played a major role in Sumerian society.

Key Terms and Places

Sumer area of Mesopotamia where the world's first civilization was developed
city-state a political unit consisting of a city and the surrounding countryside
empire land with different territories and people under a single rule
polytheism the worship of many gods
priests people who performed religious ceremonies
social hierarchy a division of society by rank or class

Section Summary

AN ADVANCED SOCIETY

In southern Mesopotamia about 3000 BC, people known as the Sumerians (soo-MER-ee-unz) created a complex, advanced society. Most people in **Sumer** (soo-muhr) lived in rural areas, but they were governed from urban areas that controlled the surrounding countryside. The size of the countryside controlled by each of these **city-states** depended on its military strength. Stronger city-states controlled larger areas. Individual city-states gained and lost power over time.

Around 2300 BC Sargon was the leader of the Akkadians (uh-KAY-dee-uhns), a people who lived to the north of Sumer. Sargon built a large army and defeated all the city-states of Sumer as well as all of northern Mesopotamia. With these conquests, Sargon established the world's first **empire**. It stretched from the Persian Gulf to the Mediterranean Sea. The Akkadian empire lasted about 150 years.

Why do you think governments are usually located in cities?

____

____

____

____

Use a world atlas to determine how many miles across the Akkadian empire extended.

Section 2, *continued*

RELIGION SHAPES SOCIETY

Religion played an important role in nearly
every aspect of Sumerian public and private life.
Sumerians practiced **polytheism**, the worship of
many gods. They believed that their gods had
enormous powers. Gods could bring a good harvest
or a disastrous flood. The gods could bring illness
or they could bring good health and wealth. The
Sumerians believed that success in every area of life
depended on pleasing the gods. Each city-state
considered one god to be its special protector.
People relied on **priests** to help them gain the gods'
favor. Priests interpreted the wishes of the gods and
made offerings to them.

A **social hierarchy** developed in Sumerian
city-states. Kings were at the top. Below them
were priests and nobles. The middle ranks included
skilled craftspeople, and merchants. Farmers and
laborers made up the large working class. Slaves
were at the bottom of the social order. Although
the role of most women was limited to the home
and raising children, some upper-class women were
educated and even became priestesses.

> **Do you think religion plays an important role in public life today? Why or why not?**
> ___________________________
> ___________________________
> ___________________________
> ___________________________

> **In Sumerian religious practice, what did priests do to try to please the gods?**
> ___________________________
> ___________________________

> **Which two groups formed the Sumerian upper classes?**
> ___________________________
> ___________________________
> ___________________________

CHALLENGE ACTIVITY

Critical Thinking: Drawing Inferences You are the king of a Sumerian
city-state. Write a letter to your priests asking them to make offerings
to the gods in order to protect your farms from a possible flood.

       Interactive Reader and Study Guide

History of the Fertile Crescent

Section 3

MAIN IDEAS

1. The Sumerians invented the world's first writing system.
2. Advances and inventions changed Sumerian lives.
3. Many types of art developed in Sumer.

Key Terms

cuneiform the Sumerian system of writing, which used symbols to represent basic parts of words

pictographs picture symbols that represented objects such as trees or animals

scribe writer

epics long poems that tell the story of a hero

architecture the science of building

ziggurat a pyramid-shaped temple tower

Section Summary

INVENTION OF WRITING

The Sumerians made one of the greatest cultural advances in history. They developed **cuneiform** (kyoo-NEE-uh-fohrm), the world's first system of writing. But Sumerians did not have pencils, pens, or paper. Instead, they used sharp reeds to make wedge-shaped symbols on clay tablets.

Sumerians first used cuneiform to keep records for business, government, and temples. As the use of cuneiform grew, simple **pictographs** evolved into more complex symbols that represented basic parts of words. Writing was taught in schools. Becoming a writer, or **scribe**, was a way to move up in social class. Scribes began to combine symbols to express complex ideas. In time, scribes wrote works on law, grammar, and mathematics. Sumerians also wrote stories, proverbs, songs, poems to celebrate military victories, and long poems called **epics**.

> Write the name of the world's first system of writing.
>
> _______________________
>
> _______________________

> What are pictographs?
>
> _______________________
>
> _______________________
>
> _______________________

 Interactive Reader and Study Guide

Section 3, *continued*

ADVANCES AND INVENTIONS

The Sumerians were the first to build wheeled vehicles like carts and wagons. They invented the potter's wheel, a device that spins wet clay as a craftsperson shapes it into bowls. They invented the ox-drawn plow and greatly improved farm production. They built sewers under city streets. They learned to use bronze to make strong tools and weapons. They named thousands of animals, plants, and minerals, and used them to produce healing drugs. The clock and the calendar we use today are based on Sumerian methods of measuring time.

> Which Sumerian invention greatly improved farm production?
>
> ______________________________
>
> ______________________________

THE ARTS OF SUMER

Sumerian remains reveal great skill in **architecture**. A pyramid-shaped **ziggurat** dominated each city. Most people lived in one-story houses with rooms arranged around a small courtyard.

> Underline the sentence that describes the kind of houses in which most Sumerians lived.

Sumerian art is renowned for sculpture and jewelry. Sculptors created statues of gods for the temples, and made small objects of ivory or rare woods. Jewelers worked with imported gold, silver, and fine stones. Earrings and other items found in the region show that Sumerian jewelers knew advanced methods for putting gold pieces together.

The Sumerians also developed a special art form called the cylinder seal. The cylinder seal was a small stone cylinder that was engraved with designs and could be rolled over wet clay to decorate containers or to "sign" documents.

Music played an important role in Sumerian society. Musicians played stringed instruments, reed pipes, drums, and tambourines both for entertainment and for special occasions.

> Name four types of musical instruments played by Sumerians.
>
> ______________________________
>
> ______________________________
>
> ______________________________
>
> ______________________________

CHALLENGE ACTIVITY

Critical Thinking: Drawing Inferences Consider the inventions of writing and the wheel. As you go through a normal day keep a list of the things you do that rely on these two inventions.

Section 4

MAIN IDEAS

1. The Babylonians conquered Mesopotamia and created a code of law.
2. Invasions of Mesopotamia changed the region's culture.
3. The Phoenicians built a trading society in the eastern Mediterranean region.

Key Terms and Places

Babylon important Mesopotamian city-state near present-day Baghdad

Hammurabi's Code the earliest known written collection of laws, comprising 282 laws that dealt with almost every part of life

chariot a wheeled, horse-drawn battle car

alphabet a set of letters that can be combined to form written words

Section Summary

THE BABYLONIANS CONQUER MESOPOTAMIA

By 1800 BC, a powerful city-state had arisen in **Babylon**, an old Sumerian city on the Euphrates. Babylon's greatest monarch (MAH-nark), Hammurabi, conquered all of Mesopotamia.

During his 42-year reign, Hammurabi oversaw many building and irrigation projects, improved the tax collection system, and brought prosperity through increased trade. He is most famous, however, for **Hammurabi's Code**, the earliest known written collection of laws. It contained laws on everything from trade, loans, and theft to injury, marriage, and murder. Some of its ideas are still found in laws today. The code was important not only for how thorough it was, but also because it was written down for all to see.

> On what river was the city of Babylon located?
> ___________________________
> ___________________________

> Why do you think it is important for laws to be written down?
> ___________________________
> ___________________________
> ___________________________

INVASIONS OF MESOPOTAMIA

Several other civilizations developed in and around the Fertile Crescent. As their armies battled each other for Mesopotamia's fertile land, control of the region passed from one empire to another. The Hittites of Asia Minor captured Babylon in 1595 BC with strong iron weapons and the skillful use of the

> Name four groups that conquered all of Mesopotamia after the Babylonians.
> ___________________________
> ___________________________
> ___________________________
> ___________________________

chariot on the battlefield. After the Hittite king was killed, the Kassites captured Babylon and ruled for almost 400 years.

The Assyrians were the next group to conquer all of Mesopotamia. They ruled from Nineveh, a city in the north. The Assyrians collected taxes, enforced laws, and raised troops through local leaders. The Assyrians also built roads to link distant parts of the empire. In 612 BC the Chaldeans, a group from the Syrian Desert, conquered the Assyrians.

Nebuchadnezzar (neb-uh-kuhd-NEZ-uhr), the most famous Chaldean king, rebuilt Babylon into a beautiful city. According to legend, his grand palace featured the famous Hanging Gardens. The Chaldeans revived Sumerian culture and made notable advances in astronomy and mathematics.

> **Which older Mesopotamian civilization did the Chaldeans admire and study?**
>
> _______________________
>
> _______________________

THE PHOENICIANS

Phoenicia, at the western end of the Fertile Crescent along the Mediterranean Sea, created a wealthy trading society. Fleets of fast Phoenician trading ships sailed throughout the Mediterrranean and even into the Atlantic Ocean, building trade networks and founding new cities. The Phoenicians' most lasting achievement, however, was the **alphabet**, a major development that has had a huge impact on the ancient world and on our own.

> **On what body of water were most Phoenician colonies located?**
>
> _______________________
>
> _______________________

CHALLENGE ACTIVITY

Critical Thinking: Drawing Inferences Make a time line with approximate dates showing the various empires and invasions that characterized the history of Mesopotamia up to the time of the Chaldeans.

Judaism and Christianity

CHAPTER SUMMARY

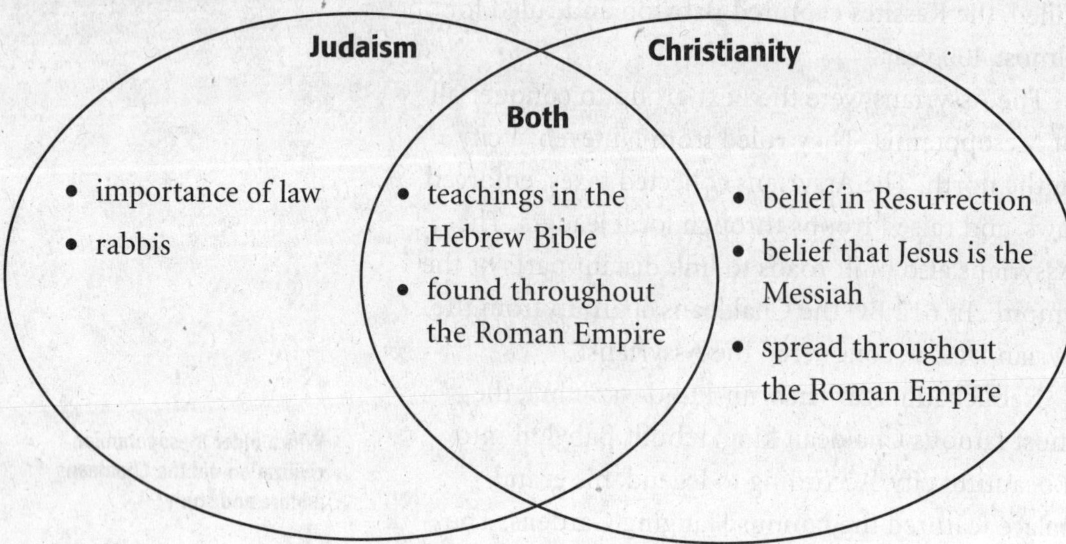

Use information from the graphic organizer to answer the following questions.

1. Recall Which religion became more widespread?

____

____

2. Interpret In which part of the diagram could *Exodus from Egypt* be added?

____

____

3. Evaluate Which of the following could NOT be placed in the overlapping area— *belief in one god, Jerusalem a holy city,* or *Gospels written by Apostles*?

____

____

Judaism and Christianity

Section 1

MAIN IDEAS

1. The Hebrews' early history began in Canaan and ended when the Romans forced them out of Israel.
2. Jewish beliefs in God, justice, and law anchor their society.
3. Jewish sacred texts describe the laws and principles of Judaism.
4. Traditions and holy days celebrate the history and religion of the Jewish people.

Key Terms and Places

Judaism the Hebrews' religion

Canaan where Abraham settled on the Mediterranean Sea

Exodus a journey of the Hebrews out of Egypt, led by Moses

monotheism the belief in one and only one god

Torah the most sacred text of Judaism

rabbis religious teachers of Judaism

Section Summary

EARLY HISTORY

The Hebrews appeared in Southwest Asia sometime between 2000 and 1500 BC. Their religion was **Judaism**. According to the Bible, the Hebrews are descended from Abraham. The Bible says that God told Abraham to lead his family to **Canaan** on the Mediterranean Sea. Later, some Hebrews moved from Canaan to Egypt.

The Hebrews were enslaved in Egypt. A leader named Moses helped the Hebrews get their freedom. He then led them on a journey out of Egypt called the **Exodus**. The Bible says that God gave Moses two stone tablets on a mountain called Sinai. A code of moral laws called the Ten Commandments was written on the tablets.

The Hebrews reached Canaan, or Israel. Israel eventually split into two kingdoms—Israel and Judah. The people of Judah became known as Jews. Invaders conquered Israel and Judah and sent the Jews out of Jerusalem as slaves. When the invaders

> Circle the name of the man who the Bible says is the ancestor of the Hebrews.

were conquered, some Jews returned home. Some moved to other places. This scattering of Jews outside of Israel is called the Diaspora.

JEWISH BELIEFS

Jews share several central beliefs. One of these is **monotheism**. Jews believe that there is one and only one God. Ideas of justice and righteousness are also important. Finally, the Jews believe in following religious and moral laws. These include those found in the Ten Commandments.

JEWISH TEXTS

Judaism has several sacred texts. These contain the religion's basic laws and principles. The **Torah**, the first part of the Hebrew Bible, is the most sacred text. The Talmud is a set of laws, commentaries, stories and folklore. Jewish **rabbis**, or religious teachers, have studied these texts for centuries.

TRADITIONS AND HOLY DAYS

There are several Jewish traditions and holy days. Hanukkah and Passover are celebrations of historical events. The two most sacred Jewish holidays are Rosh Hashanah and Yom Kippur. Rosh Hashanah celebrates the start of the new year. On Yom Kippur, Jews ask God to forgive their sins. This is the holiest day of the year for the Jews.

CHALLENGE ACTIVITY

Critical Thinking: Drawing Conclusions Imagine you are a tour guide in a museum of Jewish history. Which part of the museum do you think that tourists might enjoy the most? Write a brief recommendation.

Judaism and Christianity

Section 2

MAIN IDEAS

1. The life and death of Jesus of Nazareth inspired a new religion called Christianity.
2. Christians believe that Jesus's acts and teachings focused on love and salvation.
3. Jesus's followers taught others about Jesus's life and teachings.
4. Christianity spread throughout the Roman Empire by 400.

Key Terms and Places

Messiah a great leader the ancient Jews predicted would come to restore the greatness of Israel

Christianity a religion based on Jesus's life and teachings

Bible the holy book of Christianity

Bethlehem a small town where Jesus was born

Resurrection Jesus's rise from the dead

disciples followers

saint a person known and admired for his or her holiness

Section Summary

JESUS OF NAZARETH

Many people thought Jesus was the **Messiah**, a leader who would bring back Israel's greatness. The life and teachings of Jesus of Nazareth are the basis of a religion called **Christianity**. Stories about Jesus's life are in the **Bible**, the holy book of Christianity.

Jesus was born in the town of **Bethlehem** and spent much of his life in Nazareth. Jesus had many followers. But his teachings challenged the authority of Roman leaders. According to the Bible, they tried and executed Jesus around AD 30. Christians believe that Jesus rose from the dead. They refer to this as the **Resurrection**. They believe that Jesus next appeared to his **disciples**, or followers. He gave them instructions about how to pass on his teachings. Then he rose up to heaven.

Why was Jesus tried and executed?

____

____

What do Christians believe happened after Jesus died?

____

____

  Interactive Reader and Study Guide

Section 2, continued

JESUS'S ACTS AND TEACHINGS

According to the Bible, Jesus performed miracles. He told many parables, stories that taught lessons about how people should live. Jesus taught people to love God and love other people. Jesus also taught about salvation, or the rescue of people from sin.

Since Jesus's death, people have interpreted his teachings in different ways. As a result, different denominations, or groups, of Christianity have developed.

> Underline two of Jesus's major teachings.

JESUS'S FOLLOWERS

After Jesus's death, his followers continued to spread his teachings. The disciples Matthew, Mark, Luke, and John wrote the Gospels, which are found in the New Testament of the Bible. Paul spread Jesus's teachings throughout the Mediterranean. After his death, Paul was named a **saint**. A saint is a person known and admired for his or her holiness.

> Why do you think Paul was named a saint?
> ____________________________
> ____________________________

THE SPREAD OF CHRISTIANITY

Christianity spread quickly. Roman leaders arrested and killed some Christians who refused to worship the gods of Rome. Some emperors banned Christianity. Christians often had to worship in secret. Local leaders called bishops led each community. The bishop of Rome, or the pope, came to be viewed as the head of the Christian Church.

Christianity continued to spread throughout Rome. Then the Roman emperor Constantine converted to Christianity. He lifted the bans against the practice of the religion. Christianity eventually spread from Rome all around the world.

> Circle the name of the emperor who converted to Christianity.

CHALLENGE ACTIVITY

Critical Thinking: Understanding Cause and Effect Write a letter to Paul explaining the long-term effects of his ministry.

Judaism and Christianity

Section 3

> ### MAIN IDEAS
>
> 1. Eastern emperors ruled from Constantinople and tried but failed to reunite the whole Roman Empire.
> 2. The people of the eastern empire created a new society that was very different from society in the west.
> 3. Byzantine Christianity was different from religion in the west.

Key Terms and Places

Constantinople the eastern capital of the Roman Empire

Byzantine Empire the society that developed in the eastern Roman Empire after the west fell

mosaics pictures made with pieces of colored stone or glass

Section Summary

EMPERORS RULE FROM CONSTANTINOPLE

The capital of the Roman Empire was **Constantinople**. The city was located between two seas. It controlled trade between Asia and Europe.

The emperor Justinian ruled from 527 to 565. In an effort to reunite the old Roman empire, he sent his army to retake Italy and much of the land around the Mediterranean. Justinian made other changes as well. He simplified Roman laws and organized them into a legal system called Justinian's Code. The code helped guarantee fairer treatment for all.

> What was Justinian's Code?
> _______________________
> _______________________

Justinian had many successes, but he also made enemies. These enemies tried to overthrow him in 532. His wife Theodora convinced Justinian to stay in Constantinople and fight. With her advice, he found a way to end the riots.

> Circle the name of Justinian's wife.

The eastern empire began to decline after Justinian died. Invaders took away all the land he had gained. Nearly 900 years after Justinian died, the eastern Roman Empire finally ended. In 1453 invaders captured Constantinople. With this defeat the 1,000-year history of the eastern Roman Empire came to an end.

> Circle the date when the Roman Empire ended. Why did it end?
> _______________________

Section 3, *continued*

A NEW SOCIETY

After Justinian's death, non-Roman influences took hold throughout the empire. Many people spoke Greek, and scholars studied Greek philosophy. A new society developed. This society is called the **Byzantine Empire**. The Byzantines interacted with many groups, largely because of trade.

The eastern empire was different from the western empire in another way. Byzantine emperors had more power than western emperors did. They were the heads of the church as well as political rulers. The leaders of the church in the west were bishops and popes. The western emperors had only political power.

> What were two ways that the Byzantine Empire was different from the western empire?
>
> ___________________________
>
> ___________________________

BYZANTINE CHRISTIANITY

Christianity was central to the lives of nearly all Byzantines. Artists created beautiful works of religious art. Many Byzantine artists made **mosaics**. These were pictures made with pieces of colored stone or glass. Some were made of gold, silver, and jewels. Magnificent churches also were built.

> Underline the description of mosaics.

Over time, eastern and western Christianity became very different. People had different ideas about how to interpret and practice the religion. By the 1000s the church split in two. Eastern Christians formed the Orthodox Church.

CHALLENGE ACTIVITY

Critical Thinking: Comparing and Contrasting Create a Venn diagram illustrating the similarities and differences between the eastern and western empires. Conduct library or Internet research to find interesting details for your diagram.

History of the Islamic World

CHAPTER SUMMARY

Examples of Islamic Impact

	Western Empire
Arabia	First caliph unified Arabia for first time
Social Change	
Role of Women	
Byzantine Empire	Conquered by Ottoman Turks
Science, Medicine, and Mathematics	
Arts and Architecture	*The Thousand and One Nights*, Taj Mahal

COMPREHENSION AND CRITICAL THINKING

Use the answers to the following questions to fill in the graphic organizer above.

1. Explain Who was the first caliph? What did he accomplish?

____

____

2. Identify Cause and Effect How did the teachings of Muhammad affect early Arabian society?

____

____

3. Evaluate Name three great achievements of Islamic empires.

____

____

4. Draw a Conclusion Were the Islamic empires a good place to live if you were a woman or a non-Muslim? Explain.

____

____

History of the Islamic World

Section 1

MAIN IDEAS

1. Arabia is a mostly a desert land, where two ways of life, nomadic and sedentary, developed.

2. A new religion called Islam, founded by the prophet Muhammad, spread throughout Arabia in the 600s.

Key Terms and Places

Mecca birthplace of Muhammad

Islam religion based on messages Muhammad received from God

Muslim a person who follows Islam

Qur'an the holy book of Islam

Medina city that Muhammad and his followers moved to from Mecca in 622

mosque a building for Muslim prayer

Section Summary

LIFE IN A DESERT LAND

Arabia, in the southwest corner of Asia, is the crossroads for Africa, Europe, and Asia. Arabia is a mostly hot and dry desert of scorching temperatures and little water. Water is scarce and exists mainly in oases, or wet, fertile areas in the desert. Oases are key stops along Arabia's trade routes.

People developed two ways to live in the desert. Nomads moved from place to place. Nomads lived in tents and raised goats, sheep, and camels. They traveled with their herds to find food and water for their animals. They traveled in tribes, or groups of people. Tribe membership provided protection from danger and reduced competition for grazing lands.

Others led a settled life. Towns sprang up in oases along the trade routes. Merchants and craftspeople traded with groups of traders who traveled together in caravans. Most towns had a market or bazaar. Both nomads and caravans used these centers of trade.

Arabia was the trading crossroads for what three continents?

____

____

____

Why would a nomad prefer to travel in a tribe?

____

____

____

____

 Interactive Reader and Study Guide

A NEW RELIGION

A man named Muhammad brought a new religion to Arabia. Much of what we know about him comes from religious writings. Muhammad was born in the city of **Mecca** around 570. As a child, he traveled with his uncle's caravans. As an adult, Muhammad managed a caravan business.

Muhammad was upset that rich people did not help the poor. He often went to a cave to meditate on this problem. According to Islamic belief, when Muhammad was 40, an angel spoke to him. These messages form the basis of a religion called **Islam**. A follower of Islam is called a **Muslim**. The messages were written in the **Qur'an** (kuh-RAN), the holy book of Islam.

Muhammad taught that there was only one God, Allah. The belief in one god was a new idea for many Arabs. Before this time, Arabs prayed to many gods at shrines. The most important shrine was in Mecca. Many people traveled to Mecca every year on a pilgrimage. Muhammad also taught that the rich should give money to the poor. But rich merchants in Mecca rejected this idea.

Slowly, Muhammad's message began to influence people. The rulers in Mecca felt threatened by him. Muhammed left and went to **Medina**. His house there became the first **mosque**, or building for Muslim prayer. After years of conflict, the people of Mecca finally gave in and accepted Islam.

> Circle the name of Islam's prophet and founder.

> Where did Muhammad first hear from an angel of God?
>
> _______________________
>
> _______________________

> Why do you think the rich merchants disliked being told they should give money to the poor?
>
> _______________________
>
> _______________________
>
> _______________________
>
> _______________________

CHALLENGE ACTIVITY

Critical Thinking: Drawing Inferences If you lived in Arabia, would you choose a nomadic or more settled, sedentary life? Write a one-page description of what your life would be like.

History of the Islamic World

Section 2

> **MAIN IDEAS**
> 1. The Qur'an guides Muslims' lives.
> 2. The Sunnah tells Muslims of important duties expected of them.
> 3. Islamic law is based on the Qur'an and the Sunnah.

Key Terms

jihad literally means "to make an effort" or "to struggle"

Sunnah a collection of actions or sayings by Muhammad

Five Pillars of Islam the five acts of worship required of all Muslims

Section Summary

THE QUR'AN

After Muhammad died, his followers wrote down all of the messages he received from Allah. This collection of teachings became known as the Qur'an. Muslims believe that the Qur'an is the exact word of God as it was told to Muhammad. Like the Jewish and Christian bibles, the Qur'an says there is one God (Allah). Islam teaches that there is a definite beginning and end to the world. On that final day, Muslims believe, God will judge all people. Those who have obeyed God's orders will be granted life in paradise. Those who have not obeyed God will be punished.

> Circle the name of Islam's most important holy book.

Muslims believe that God wishes them to follow many rules in order to be judged a good person. These rules affect the everyday life of Muslims. In the early days of Islam, these rules led to great changes in Arabian society. For example, owning slaves was forbidden.

Jihad (ji-HAHD) is an important Islamic concept. Literally, jihad means "to make an effort" or "to struggle." It refers to the internal struggle of a Muslim trying to follow Islamic beliefs. It can also mean the struggle to defend the Muslim community

> How do you think "jihad" came to mean "holy war?"
>
> _______________________
>
> _______________________
>
> _______________________
>
> _______________________

Section 2, continued

or convert people to Islam. The word has also been translated as "holy war."

THE SUNNAH

Another important holy book in Islam is the **Sunnah** (SOOH-nuh), a collection of Muhammad's words and actions. The Sunnah spells out the main duties for Muslims. These are known as the **Five Pillars of Islam**. The first pillar is a statement of faith. The second pillar says a Muslim must pray five times daily. The third pillar is a yearly donation to charity. The fourth pillar is fasting during the holy month of Ramadan (RAH-muh-dahn). The fifth pillar is the hajj (HAJ), a pilgrimage to Mecca. The hajj must be made at least once in a lifetime.

The Sunnah also preaches moral duties that must be met in daily life, in business, and in government. For example, it is considered immoral to owe someone money or to disobey a leader.

ISLAMIC LAW

The Qur'an and the Sunnah form the basis of Islamic law, or Shariah (shuh-REE-uh). Shariah lists rewards or punishments for obeying or disobeying laws. Shariah punishments can be severe. Shariah makes no distinction between religious and secular life. Most Islamic countries today blend Islamic law with a legal system much like that in the United States.

CHALLENGE ACTIVITY

Critical Thinking: Drawing Inferences Write a brief essay evaluating the differences and similarities between the two earlier religions of Judaism and Christianity with Islam. Focus not only on beliefs but also on practices and what social conditions might have influenced these practices in all the religions.

History of the Islamic World

Section 3

MAIN IDEAS

1. Muslim armies conquered many lands into which Islam slowly spread.
2. Trade helped Islam spread into new areas.
3. Three Muslim empires controlled much of Europe, Asia, and Africa from the 1400s to the 1800s.

Key Terms and Places

caliph title of the highest Islamic leader

tolerance acceptance

Baghdad city that became the capital of the Islamic Empire in 762

Córdoba Muslim city in Spain that became the largest and most advanced city in western Europe in the early 900s

janissaries slave boys converted to Islam and trained as soldiers

Istanbul capital of the Ottoman Empire; formerly Constantinople

Esfahan capital of the Safavid Empire

Section Summary

MUSLIM ARMIES CONQUER MANY LANDS

After Muhammad's death Abu Bakr (UH-boo BAK-uhr) was the leader of Islam. He was the first **caliph** (KAY-luhf). This title was used for the highest Islamic leader. Abu Bakr unified Arabia. The Arab army conquered the Persian and Byzantine empires.

Later caliphs conquered lands in Central Asia, northern India, and North Africa. They controlled eastern Mediterranean trade routes. After many years of fighting, the Berbers of North Africa converted to Islam. A combined Arab and Berber army conquered Spain and ruled for 700 years.

> What present-day countries mark the eastern and western boundaries of the Islamic empire?
>
> _______________________
>
> _______________________
>
> _______________________

TRADE HELPS ISLAM SPREAD

Arab merchants took Islamic beliefs and practices with them to new lands. Coastal trading cities developed into large Muslim communities.

Muslims generally practiced **tolerance**, or acceptance. They did not ban all other religions in

> Why do you think trade flourishes in coastal cities?
>
> _______________________
>
> _______________________
>
> _______________________
>
> _______________________

Section 3, *continued*

their lands. More people began speaking Arabic and practicing Islam. The Arabs also took on non-Muslim customs. Cultural blending changed Islam into a religion of many cultures. The development of Muslim cities like **Baghdad** and **Córdoba** reflected this blending of cultures.

THREE MUSLIM EMPIRES

In the 1200s, Muslim Turks known as Ottomans attacked the Byzantine Empire. They trained **janissaries**, boys from conquered towns who were enslaved and converted to Islam. The janissaries fought fiercely. In 1453 the Ottomans took Constantinople and renamed it **Istanbul**. This ended the Byzantine Empire. By 1566 the Ottomans took control of the eastern Mediterranean and parts of Europe.

The Safavids (sah-FAH-vuhds) gained power in the east. They soon came into conflict with the Ottomans. The conflict stemmed from an old disagreement about who should be caliph. In the mid-600s, Islam had split into two groups—the Sunni and the Shia. The Ottomans were Sunni, and the Safavids were Shia. The Safavid Empire conquered Persia in 1501 and soon grew wealthy, building glorious mosques in **Esfahan**, their capital.

East of the Safavid Empire, in India, lay the Mughal (MOO-guhl) Empire. The Mughals united many diverse peoples and were known for their architecture—particularly the Taj Mahal. Under the leader Akbar, the Mughal Empire was known for its religious tolerance. But more restrictive policies after his death led to the end of the empire.

> Underline the phrase that tells where the Ottomans found fierce soldiers to fight in their armies.

> What date signifies the end of the Byzantine Empire?
>
> _______________________
>
> _______________________

> Think of what you know about the Middle East today. Does the conflict between the Sunnis and the Shias continue?
>
> _______________________
>
> _______________________
>
> _______________________
>
> _______________________

CHALLENGE ACTIVITY

Critical Thinking: Drawing Inferences Draw a time line marking the major Muslim conquests and a map to show the size of Islamic territory.

Interactive Reader and Study Guide

Section 4

MAIN IDEAS

1. Muslim scholars made lasting contributions to the fields of science and philosophy.
2. In literature and the arts, Muslim achievements included beautiful poetry, memorable short stories, and splendid architecture.

Key Terms

Sufism a movement of Islam, based on the belief that one must have a personal relationship with God

minarets tall towers on mosques from which Muslims are called to prayer

calligraphy decorative writing

Section Summary

SCIENCE AND PHILOSOPHY

Islamic scholars made great advances in many fields. These included astronomy, geography, math, and science. At Baghdad and Córdoba, Greek and other writings were translated into Arabic. A common language helped scholars share research.

Muslim scientists built observatories to study the stars. They also improved the astrolabe. The Greeks had invented this tool to chart the position of the stars. The astrolabe would later be used in sea exploration.

It was a Muslim mathematician who invented algebra. Muslims found better ways to calculate distance and make precise maps. They also used the stars to navigate. Muslim merchants and explorers traveled far and wide. One great explorer was Ibn Battutah. He traveled to Africa, India, China, and Spain.

Muslims were also known in medicine. They added greatly to Greek and Indian medicine. Muslims also started the first school of pharmacy.

> **Which two cities came to be recognized as the cultural capitals of Islam during the Middle Ages?**
>
> _____________________
>
> _____________________

> **Why do you think the astrolabe would be useful in sea exploration?**
>
> _____________________
>
> _____________________
>
> _____________________

Section 4, *continued*

A doctor in Baghdad found out how to detect and treat the disease smallpox. Another doctor, known in the West as Avicenna (av-uh-SEN-uh), wrote a medical encyclopedia. It was used widely in Europe for centuries.

A new philosophy developed. It was called **Sufism** (SOO-fi-zuhm). People who practice Sufism are Sufis (SOO-feez). Sufis seek a personal relationship with God. Sufism has brought many followers to Islam.

LITERATURE AND THE ARTS

Poetry and short stories were popular among Muslims. The collection of stories called *The Thousand and One Nights* is still one of the best-loved books in the world. Sufi poets were popular, including the famous Omar Khayyám (OH-mahr ky-AHM).

There were many achievements in architecture. Rulers liked to be patrons. Patrons helped fund the design and construction of beautiful mosques. The main part of a mosque is a huge hall where thousands of people gather to pray. Often mosques have large domes and **minarets**.

Islam does not allow artists to show animals or humans in religious art. Muslims believe only Allah can create humans and animals or their images. In part for this reason, Muslim artists turned to **calligraphy**. This decorative writing became an art form.

CHALLENGE ACTIVITY

Critical Thinking: Drawing Inferences Islamic culture made many advances in science, medicine, and art that still affect us today. Pick the advance that you think is the most important to our modern society, and write a one-page paper explaining your position.

> **What two advances in medicine were made by Muslim doctors?**
> ___________________
> ___________________
> ___________________
> ___________________

> **What is the name of Islam's great collection of stories?**
> ___________________
> ___________________
> ___________________

> **Underline the sentence that helps to explain why Muslim artists developed calligraphy as a fine art.**

Interactive Reader and Study Guide

The Eastern Mediterranean

CHAPTER SUMMARY

The Eastern Mediterranean

	Capital	History	People/Culture	Government	Economy
Turkey					
Israel					
Syria					
Lebanon					
Jordan					

COMPREHENSION AND CRITICAL THINKING

Use the graphic organizer to answer the following questions.

1. Identify In the chart above, where would you list details about civil war in Lebanon? Where would you record population statistics about Israel?

____

____

2. Interpret Of the following seven words (*Beirut, Amman, Damascus, basalt, Jerusalem, kibbutz, Ankara*), which two do *not* belong in the category "Capital"?

____

____

3. Sequence If you were going to describe the countries by size, from smallest to largest, in what order would you describe them?

____

____

____

The Eastern Mediterranean

Section 1

MAIN IDEAS

1. The Eastern Mediterranean's physical features include the Bosporus, the Dead Sea, rivers, mountains, deserts, and plains.
2. The region's climate is mostly dry with little vegetation.
3. Important natural resources in the Eastern Mediterranean include valuable minerals and the availability of water.

Key Terms and Places

Dardanelles body of water that connects the Sea of Marmara and the Mediterranean Sea; part of the narrow waterway that separates Europe and Asia

Bosporus body of water that connects the Black Sea and the Sea of Marmara; part of the narrow waterway that separates Europe and Asia

Jordan River river that begins in Syria and flows south through Israel and Jordan, finally emptying into the Dead Sea

Dead Sea lowest point on any continent and the world's saltiest body of water

Syrian Desert a desert of rock and gravel covering much of Syria and Jordan

Section Summary

PHYSICAL FEATURES

The Eastern Mediterranean is part of a larger region called Southwest Asia, or the Middle East. The **Dardanelles,** the **Bosporus,** and the Sea of Marmara separate Europe from Asia. A small part of Turkey lies in Europe. The larger Asian part of Turkey is called Anatolia.

The **Jordan River** flows from Syria to Israel and Jordan, then empties into the **Dead Sea,** the world's saltiest body of water.

Two mountain systems stretch across Turkey. The Pontic Mountains lie in the north, and the Taurus Mountains lie in the south. A narrow plain runs from Turkey into Syria. The Euphrates River flows southeast through this plain. Hills, valleys, and plateaus are located farther inland. Two mountain ridges run north-south. One runs from Syria through western Jordan. The other runs through Lebanon and Israel.

> **What three bodies of water separate Europe and Asia?**
> _______________________
> _______________________
> _______________________

> **Which two mountain systems stretch across Turkey?**
> _______________________
> _______________________

Section 1, *continued*

CLIMATE AND VEGETATION

The Eastern Mediterranean is a mostly dry region. However, there are important variations. Turkey's Black Sea coast and the Mediterranean coast to northern Israel have a Mediterranean climate. Central Syria and lands farther south have a desert climate. Much of Turkey has a steppe climate, and a small area in the northeast has a humid subtropical climate.

The driest areas are the deserts. The **Syrian Desert** covers much of Syria and Jordan. The Negev Desert lies in southern Israel.

> Circle the words and phrases that describe some of the different climates in the eastern Mediterranean.

NATURAL RESOURCES

Because the region is so dry, water is a valuable resource. Commercial farming relies on irrigation. Subsistence farming and herding takes place in drier areas.

Many minerals, including sulfur, mercury, and copper, are found in the region. Phosphates are produced in Syria, Jordan, and Israel. They are used to make fertilizers. The area also exports asphalt, the dark tarlike material used to pave streets.

> What mineral resources are found in the region?
>
> _______________________
> _______________________
> _______________________
> _______________________

CHALLENGE ACTIVITY

Critical Thinking: Drawing Inferences Based on what you've learned about the climates in the Eastern Mediterranean region, write an essay describing which location you think would be best for farming and why.

Interactive Reader and Study Guide

The Eastern Mediterranean

Section 2

MAIN IDEAS

1. Turkey's history includes invasion by the Romans, rule by the Ottomans, and a twentieth-century democracy.
2. Turkey's people are mostly ethnic Turks, and its culture is a mixture of modern and traditional.
3. Today, Turkey is a democratic nation seeking economic opportunities as a future member of the European Union.

Key Terms and Places

Ankara the capital of Turkey

Istanbul Turkey's largest city

secular religion is kept separate from government

Section Summary

HISTORY

About 8,000 years ago the area that is now Turkey was home to the world's earliest farming villages.

Turkey has been invaded by powerful empires for centuries. The Romans were the first empire to invade the area. They captured Byzantium, located between Europe and Asia, and renamed it Constantinople. After the fall of Rome, Constantinople became the capital of the Byzantine Empire.

Seljuk Turks, a nomadic people from Central Asia, invaded the area in the AD 1000s. In 1453 the Ottoman Turks captured Constantinople and made it the capital of the Islamic Empire. The Ottoman Empire was very powerful during the 1500s and 1600s, controlling territory in northern Africa, southwestern Asia, and southeastern Europe. The Ottomans fought on the losing side of World War I, and lost most of their territory at the end of the war.

> **What did the Romans rename Byzantium?**
>
> _______________________
>
> _______________________

> **Who were the Seljuk Turks?**
>
> _______________________
>
> _______________________

     Interactive Reader and Study Guide

Section 2, *continued*

Military officers took over the government after World War I, led by Mustafa Kemal. He later adopted the name Kemal Atatürk, which means Father of Turks. Atatürk created the democratic nation of Turkey and moved the capital to **Ankara** from Constantinople, which was renamed **Istanbul**.

Atatürk believed in modernizing Turkey, mainly by adopting some Western methods. He banned certain types of traditional clothing of both men and women, made new laws allowing women to vote and hold office, replaced the Arabic alphabet with the Latin alphabet, and adopted the metric system.

> Who was the leader of Turkey after World War I?
>
> ___________________

> Why did Atatürk make so many changes and laws?
>
> ___________________
> ___________________

PEOPLE AND CULTURE

Most of the people living in Turkey are ethnic Turks. Kurds are the largest minority, making up 20 percent of the population.

Turkey's culture today reflects Kemal Atatürk's changes. He created a cultural split between the urban middle class and rural villagers. In general, middle-class lifestyle and attitude reflect middle-class Europeans, while rural Turks are more traditional and reflect Islamic influences.

> What is the largest minority living in Turkey today?
>
> ___________________
> ___________________
> ___________________

TURKEY TODAY

Istanbul is Turkey's largest city, but the government meets in the capital of Ankara. Turkey has a legislature called the National Assembly. A president and prime minister share executive power. Although most of the people living in Turkey are Muslim, Turkey is a **secular** state.

The economy in Turkey is based on important industries, including textiles and clothing, cement, and electronics, as well as agriculture.

> What is the National Assembly?
>
> ___________________
> ___________________

CHALLENGE ACTIVITY

Critical Thinking: Drawing Inferences In what ways do you think Turkey might benefit by becoming a member of the European Union?

 Interactive Reader and Study Guide

The Eastern Mediterranean

Section 3

MAIN IDEAS

1. Israel's history includes the ancient Hebrews and the creation of the nation of Israel.
2. In Israel today, Jewish culture is a major part of daily life.
3. The Palestinian Territories are areas within Israel controlled partly by Palestinian Arabs.

Key Terms and Places

Diaspora the scattering of the Jewish population

Jerusalem the capital of Israel

Zionism a movement that called for Jews to establish a country or community in Palestine

kosher the term used to refer to Jewish dietary laws

kibbutz a large farm where people share everything in common

Gaza a small, crowded piece of coastal land disputed over by Jews and Arabs

West Bank a largely populated, rural piece of land disputed over by Jews and Arabs

Section Summary

HISTORY

Israel is often referred to as the Holy Land. It is home to sacred sites for Jews, Muslims, and Christians. The Hebrews established the kingdom of Israel about 3,000 years ago. In the 60s BC the Roman Empire conquered the region, calling it Palestine, and forced most Jews to leave the region. This was known as the **Diaspora**. Arabs then conquered the land, but it was later invaded by Christian Crusaders, who captured the city of **Jerusalem**, but were eventually driven out. Palestine was part of the Ottoman Empire until it came under British control after World War I.

In the late 1800s European Jews began a movement called **Zionism** that called for Jews to establish a country or community in Palestine. After World War II, Jewish leaders declared Palestine the nation of Israel. Arabs living in the region opposed

> **Why is Israel often referred to as the Holy Land?**
>
> _______________________
>
> _______________________

> **When did the Roman Empire conquer the region?**
>
> _______________________

the new nation. Israel and Arab countries have fought in serveral wars over this issue, and disputes between the two sides continue today.

ISRAEL TODAY

Despite its problems, today Israel is a modern, democratic country with a diverse economy. About 80 percent of Israel's population is Jewish. The rest of the population is mostly Arab. Tel Aviv is Israel's largest city.

> **What percentage of Israel's population is Jewish?**
>
> ___________________

Jewish holidays and traditions are an important aspect of Israeli Jewish culture. Many Jews follow a **kosher** diet based on ancient religious laws. About 100,000 Israeli Jews live in **kibbutzim**, large farms where people share everything in common.

> **What are kibbutzim?**
>
> ___________________
>
> ___________________

THE PALESTINIAN TERRITORIES

In 1967 Israel captured land occupied by Palestinian Arabs—**Gaza**, the **West Bank**, and East Jerusalem. Since then Jews and Arabs have fought over the right to live in these two regions.

> **Which areas of land have been the source of the greatest conflict, tension, and violence between Arabs and Israelis?**
>
> ___________________
>
> ___________________

In the 1990s Israel agreed to turn over parts of the territories to Palestinians if the Palestinian leadership—the Palestinian Authority—agreed to work for peace. In 2005, Israelis transferred Gaza to the Palestinian Authority. The future of the peace process remains uncertain.

CHALLENGE ACTIVITY

Critical Thinking: Drawing Inferences Based on what you've learned about the conflicts between Israelis and Arabs, write a proposal you think would help ensure lasting peace. Be sure to explain why you think your proposal would be successful.

The Eastern Mediterranean

Section 4

MAIN IDEAS

1. Syria, once part of the Ottoman Empire, is an Arab country ruled by a powerful family.
2. Lebanon is recovering from civil war and its people are divided by religion.
3. Jordan has few resources and is home to Bedouins and Palestinian refugees.

Key Terms and Places

Damascus the capital of Syria

Beirut the capital of Lebanon

Bedouins Arab-speaking nomads who mostly live in the deserts of Southwest Asia

Amman the capital of Jordan

Section Summary

SYRIA

The capital of Syria, **Damascus**, is believed to be the oldest continuously inhabited city in the world. Syria became part of the Ottoman Empire in the 1500s. After World War I, France controlled Syria. Syria gained independence in the 1940s.

The Syrian government was led by Hafiz al-Assad from 1971 to 2000. Assad's son, Bashar, was elected president after his father's death in 2000. Syria's government owns the country's oil refineries, larger electrical plants, railroads, and some factories.

More than 18 million people live in Syria. About 90 percent of the population is Arab, and the remaining 10 percent include Kurds and Armenians. About 74 percent of Syrians are Sunni Muslim, about 16 percent are Alawites and Druze, and about 10 percent are Christian. There are also small Jewish communities in some Syrian cities.

LEBANON

Lebanon is a small, mountainous country. Many ethnic minority groups settled in Lebanon during the Ottoman Empire. After World War I it was

> Which country controlled Syria after World War I?
>
> _______________________

> Underline the sentence that describes Syria's main industries and who owns them.

     Interactive Reader and Study Guide

Section 4, *continued*

controlled by France. Lebanon finally gained its
independence in the 1940s.

> When did Lebanon gain its independence?

Most Lebanese people are Arab, but they are
divided by religion. The main religions in Lebanon
are Islam and Christianity, with each of these
groups divided into smaller groups. Muslims are
divided into Sunni, Shia, and Druze. The Maronites
are the largest Christian group.

> What are the two main religious groups in Lebanon?

After gaining independence, Christian and
Muslim politicians shared power. However, over
time this cooperation broke down and tensions
mounted. Warfare between the groups lasted until
1990. The capital, **Beirut**, was badly damaged.

JORDAN

The country of Jordan was created after World
War I. The British controlled the area until the
1940s, when the country gained full independence.
King Hussein ruled Jordan from 1952 to 1999. He
enacted some democratic reforms in the 1990s.

> When was Jordan created?

Jordan is a poor country with limited resources.
Many people in Jordan are **Bedouins**, or Arab-
speaking nomads who live mainly in the deserts
of Southwest Asia. **Amman**, the capital, is Jordan's
largest city. The country's resources include
phosphates, cement, and potash. In addition,
the tourism and banking industries are growing.
Jordan depends on economic aid from oil-rich
Arab nations and the United States.

> Underline the sentences that describe Jordan's economy.

CHALLENGE ACTIVITY

Critical Thinking: Drawing Inferences Based on what you've learned
about Syria, Lebanon, and Jordan, how do you think each country
could work to maintain peace and improve their economies? Why do
you think your plan would work?

                                                    Interactive Reader and Study Guide

The Arabian Peninsula, Iraq, and Iran

CHAPTER SUMMARY

The Arabian Peninsula, Iraq, and Iran make up a mostly desert region with very valuable oil resources.

- Major physical features of the Arabian Peninsula, Iraq, and Iran are desert plains and mountains.
- The region has a dry climate and little vegetation.
- Most of the world is dependent on oil, a resource that is exported from this region.

Most countries of the Arabian Peninsula share three main characteristics: Islamic religion and culture, monarchy as a form of government, and valuable oil resources.

- Islamic culture and an economy greatly based on oil influence life in Saudi Arabia.
- Most other Arabian Peninsula countries are monarchies influenced by Islamic culture and oil resources.

Iraq, a country with a rich culture and natural resources, faces the challenge of rebuilding after years of conflict.

- Iraq's history includes rule by many conquerors and cultures, as well as recent conflicts and wars.
- Most people in Iraq are Arabs, but Kurds live in the north.
- Iraq today must rebuild its government and economy, which have suffered years of conflict.

Islam is a huge influence on government and daily life in Iran.

- Iran's history includes great empires and an Islamic republic.
- In Iran today, Islamic religious leaders restrict the rights of most Iranians.

COMPREHENSION AND CRITICAL THINKING

Use information from the graphic organizer to answer the following questions.

1. Recall What do most countries of the Arabian Peninsula have in common?

____

____

2. Sequence How has the type of rule in Iran changed over time?

____

____

____

The Arabian Peninsula, Iraq, and Iran

Section 1

MAIN IDEAS

1. Major physical features of the Arabian Peninsula, Iraq, and Iran are desert plains and mountains.
2. The region has a dry climate and little vegetation.
3. Most of the world is dependent on oil, a resource that is exported from this region.

Key Terms and Places

Arabian Peninsula region of the world that has the largest sand desert in the world

Persian Gulf body of water surrounded by the Arabian Peninsula, Iran, and Iraq

Tigris River river that flows across a low, flat plain in Iraq and joins the Euphrates River

Euphrates River river that flows across a low, flat plain in Iraq and joins the Tigris River

oasis a wet, fertile area in a desert that forms where underground water bubbles to the surface

wadis dry streambeds

fossil water water that is not being replaced by rainfall

Section Summary

PHYSICAL FEATURES

The region of the **Arabian Peninsula**, Iraq, and Iran has huge deserts. Not all deserts are sand. Some are bare rock or gravel. The region forms a semicircle, with the **Persian Gulf** at the center.

> Underline the words that tell you what deserts can be made of.

The region's main landforms are rivers, plains, plateaus, and mountains. The two major rivers are the **Tigris** and **Euphrates** in Iraq. They make a narrow area good for crops. This area was called Mesopotamia in ancient times.

> List the four main landforms of this region.
>
> ________________________
>
> ________________________
>
> ________________________

The Arabian Peninsula has flat, open plains in the east. In the south, desert plains are covered with sand. Deserts in the north are covered with volcanic rock. The peninsula rises slowly towards the Red Sea. This makes a high landscape of mountains and flat plateaus. The highest point is in the mountains in Yemen. Plateaus and mountains also cover most of Iran.

CLIMATE AND VEGETATION

This region has a desert climate. It can get very hot in the day and very cold at night. The Rub' al-Khali desert in Saudi Arabia is the world's largest sand desert. Its name means "Empty Quarter," because it has so little life.

Some areas with plateaus and mountains get rain or snow in winter. Some mountain peaks get more than 50 inches of rain a year. Trees grow in these areas. They also grow in **oases** in the desert. At an oasis, underground water bubbles up. Some plants also grow in parts of the desert. Their roots either go very deep or spread out very far to get as much water as they can.

RESOURCES

Water is one of this region's two most valuable resources. But water is scarce. Some places in the desert have springs that give water. Wells also provide water. Some wells are dug into dry streambeds called **wadis**. Other wells go very deep underground. These often get **fossil water**. This is water that is not replaced by rain, so these wells will run dry over time.

Oil is the region's other important resource. This resource is plentiful. Oil has brought wealth to the countries that have oil fields. But oil cannot be replaced once it is taken. Too much drilling for oil may cause problems in the future.

CHALLENGE ACTIVITY

Critical Thinking: Designing Design an illustrated poster using the term *Persian Gulf*. For each letter, write a word containing that letter that tells something about the region.

> Is the desert always hot? Explain your answer.
> ___________________
> ___________________
> ___________________
> ___________________

> Underline the sentence that explains how desert plants get water.

> Circle two important resources in this region.

Interactive Reader and Study Guide

The Arabian Peninsula, Iraq, and Iran

Section 2

MAIN IDEAS

1. Islamic culture and an economy greatly based on oil influence life in Saudi Arabia.
2. Most other Arabian Peninsula countries are monarchies influenced by Islamic culture and oil resources.

Key Terms

Shia branch of Islam in which Muslims believe that true interpretations of Islamic teachings can only come from certain religious and political leaders

Sunni branch of Islam in which Muslims believe in the ability of the majority of the community to interpret Islamic teachings

OPEC Organization of Petroleum Exporting Countries, an international organization whose members work to influence the price of oil on world markets by controlling the supply

Section Summary

SAUDI ARABIA

Saudi Arabia is the largest country on the Arabian Peninsula. It is a major center of religion and culture. Its economy is one of the strongest in this region.

Most Saudis speak Arabic. Islam is a strong influence on their culture and customs. This religion was started in Saudi Arabia by Muhammad. Most Saudis follow one of two branches of Islam—**Shia** or **Sunni**. About 85 percent of Saudi Muslims are Sunni.

Who started Islam and where?

____

____

Islam influences Saudi culture in many ways. It teaches modesty, so traditional clothing is long and loose, covering the arms and legs. Women rarely go out in public without a husband or male relative. But women can own and run businesses.

The country's government is a monarchy. There are also local officials. Only men can vote.

Section 2, *continued*

Saudi Arabia is an important member of **OPEC**, an organization with members from different countries. OPEC works to control oil supplies to influence world oil prices.

Saudi Arabia also has challenges. It has very little freshwater to grow crops, so it has to import most of its food. There is high unemployment. This is partly because young people choose to study religion instead of other subjects the economy needs.

OTHER COUNTRIES OF THE ARABIAN PENINSULA

There are six smaller countries of this region: Kuwait, Bahrain, Qatar, the United Arab Emirates (UAE), Oman, and Yemen. Like Saudi Arabia, they are all influenced by Islam and most have monarchies and depend on oil.

Most of these countries are rich. Yemen is the poorest. Oil was only discovered there in the 1980s.

Most of these countries have a monarchy. Some also have elected officials. Yemen's government is elected, but political corruption has been a problem.

Some of these countries support their economy in other ways besides oil. Bahrain's oil began to run out in the 1990s. Banking and tourism are now important. Qatar and the UAE also have natural gas. Oman does not have as much oil as other countries, so it is trying to create new industries.

The Persian Gulf War started in 1990 when Iraq invaded Kuwait. The United States and other countries helped Kuwait defeat Iraq. Many of Kuwait's oilfields were destroyed during the war.

CHALLENGE ACTIVITY

Critical Thinking: Developing Imagine you are an official in one of the countries of the Arabian Peninsula and oil is running low. You need to develop a plan for new ways to support the economy. Based on what you know about the region's physical and human geography, list several strategies and state your reasons for why they might work.

> Underline two challenges Saudi Arabia faces.

> List the six smaller countries on the Arabian Peninsula.

> Circle three ways besides oil that countries in this region are supporting their economies.

 Interactive Reader and Study Guide

The Arabian Peninsula, Iraq, and Iran

Section 3

MAIN IDEAS

1. Iraq's history includes rule by many conquerors and cultures, as well as recent conflicts and wars.
2. Most people in Iraq are Arabs, but Kurds live in the north.
3. Iraq today must rebuild its government and economy, which have suffered years of conflict.

Key Terms and Places

embargo limit on trade

Baghdad capital of Iraq

Section Summary

HISTORY

The world's first civilization was in Iraq, in the area called Mesopotamia. Throughout history, many cultures and empires conquered Mesopotamia, including Great Britain in World War I. In the 1950s, Iraqi army officers overthrew British rule.

> Circle the location of the world's first civilization.

Iraq's recent history includes wars and a harsh, corrupt leader. In 1968, the Ba'ath Party took power. In 1979 the party's leader, Saddam Hussein, became president. He restricted the press and people's freedoms. He killed an unknown number of political enemies.

> Why was Saddam Hussein considered a harsh leader?
>
> _______________________
> _______________________
> _______________________
> _______________________

Saddam led Iraq into two wars. In 1980, Iraq invaded Iran. The Iran-Iraq War lasted until 1988. In 1990, Iraq invaded Kuwait. Western leaders worried about Iraq controlling too much oil and having weapons of mass destruction. An alliance of countries led by the United States forced Iraq from Kuwait. After the war, Saddam Hussein did not accept all the terms of peace, so the United Nations placed an **embargo** on Iraq. This hurt the economy.

> Underline the word that means a limit on trade.

After the terrorist attacks of September 11, 2001, the U.S. government officials believed that Iraq aided terrorists. In March 2003, the United States invaded Iraq. Saddam Hussein went into hiding and Iraq's government fell. Saddam Hussein was later found and arrested.

> **Why did the U.S. government invade Iraq?**
> _______________________
> _______________________
> _______________________
> _______________________

PEOPLE AND CULTURE

Most of Iraq's people belong to two ethnic groups. The majority are Arab, who speak Arabic. The others are Kurd, who speak Kurdish in addition to Arabic. Kurds live in a large region in the north of Iraq.

Most people of Iraq are Muslim. About 60 percent are Shia Muslims and live in the south. About one third are Sunni Muslims and live in the north.

> **Circle the ethnic group that most people in Iraq belong to.**

IRAQ TODAY

Today Iraq is slowly recovering from war. **Baghdad**, Iraq's capital of 8 million people, was badly damaged. People lost electricity and running water. After the war, the U.S. military and private companies helped to restore water and electricity, and to rebuild homes, businesses, and schools.

In January 2005 the people of Iraq took part in democracy for the first time. They voted for members of the National Assembly. This group's main task was to write Iraq's new constitution.

Iraq is trying to rebuild a strong economy. Oil and crops are important resources. It may take years for Iraq to rebuild structures such as schools, hospitals, and roads. It may be even harder to create a free society and strong economy.

> **The U.S. census for the year 2000 reports that about 3,700,000 people live in the city of Los Angeles. How does the population of Baghdad compare with this?**
> _______________________
> _______________________
> _______________________

> **Underline the main task of Iraq's National Assembly.**

CHALLENGE ACTIVITY

Critical Thinking: Analyzing Think about the task of writing a constitution for a whole country. Based on what you know about Iraq, create an outline for the nation's constitution that addresses issues that you think will be important to the new government.

The Arabian Peninsula, Iraq, and Iran

Section 4

MAIN IDEAS

1. Iran's history includes great empires and an Islamic republic.
2. In Iran today, Islamic religious leaders restrict the rights of most Iranians.

Key Terms and Places

shah king

revolution a drastic change in a country's government and way of life

Tehran capital of Iran

theocracy a government ruled by religious leaders

Section Summary

HISTORY

Iran today is an Islamic republic. In the past, the region was ruled by the Persian Empire and a series of Muslim empires.

The Persian Empire was a great center of art and learning. It was known for architecture and many other arts, including carpets. The capital, Persepolis, had walls and statues that glittered with gold, silver, and jewels. When Muslims conquered the region, they converted the Persians to Islam. But most people kept their Persian culture.

In 1921 an Iranian military officer took charge. He claimed the Persian title of **shah**, or king. In 1941 his son took control. This shah was an ally of the United States and Britain. He tried to make Iran more modern, but his programs were not popular.

In 1979 Iranians began a **revolution**. They overthrew the shah. The new government set up an Islamic republic that follows strict Islamic law.

Soon after the revolution began, Iran's relations with the United States broke down. A mob of students attacked the U.S. Embassy in **Tehran**. Over 50 Americans were held hostage for a year.

> Why do you think most people kept their Persian culture after Muslims conquered the region?
>
> _______________________
> _______________________
> _______________________
> _______________________

> Underline the phrase that explains what an Islamic republic does.

IRAN TODAY

Iran is unique in Southwest Asia, where most people are Arabs and speak Arabic. In Iran, more than half the people are Persian. They speak Farsi, the Persian language.

Iran has one of Southwest Asia's largest populations. It has about 68 million people, and over 35 million are younger than 25.

Iran is very diverse. Along with Persians, Iranian ethnic groups include Azerbaijanis, Kurds, Arabs, and Turks. Most Iranians are Shia Muslim. About 10 percent are Sunni Muslim. Others practice Christianity, Judaism, and other religions.

Persian culture influences life in Iran in many ways. People celebrate the Persian New Year, Nowruz. Persian food is an important part of most family gatherings.

Iran's economy is based on oil. There are also other industries, including carpet production and agriculture.

Iran's government is a **theocracy**. Its rulers, or *ayatollahs,* are religious leaders. The country also has an elected president and parliament.

Iran's government has supported many hard-line policies, such as terrorism. Today, the United States and other nations are concerned about Iran's nuclear program as a threat to world security.

> **Estimate the percentage of Iran's population that is under age 25.**
>
> ________________________

> **Circle two ways that Persian culture is part of people's lives today.**

CHALLENGE ACTIVITY

Critical Thinking: Making Inferences Imagine that you are living in Persepolis at the time the Muslims conquered the Persian Empire. Write a diary entry about what is changing in your life, what is staying the same, and how you feel about what is happening.

Central Asia

CHAPTER SUMMARY

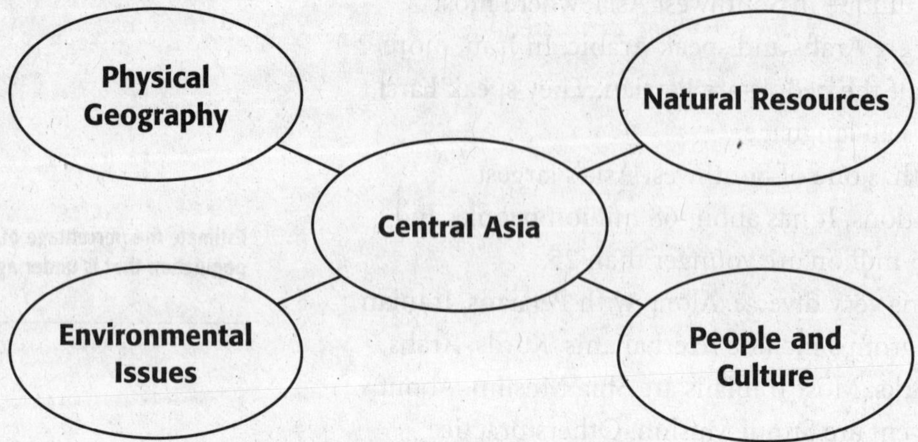

COMPREHENSION AND CRITICAL THINKING

Use the graphic organizer to answer the following questions.

1. **Identify** Of the following five choices, which three are the main natural resources of Central Asia: *oil and gas, cotton, minerals, water, trees*? Draw three lines extending from the "Natural Resources" circle and write the answers on the lines you drew.

2. **Summarize** What features characterize Central Asia's physical geography?

____

____

3. **Analyze** For each of the following words or phrases, draw a line extending from the item it belongs with and then write the word or phrase on the line you drew: *gold, shrinking Aral Sea, yurts*.

Central Asia

Section 1

MAIN IDEAS

1. Key physical features of landlocked Central Asia include rugged mountains.
2. Central Asia has a harsh, dry climate that makes it difficult for vegetation to grow.
3. Key natural resources in Central Asia include water, oil and gas, and minerals.

Key Terms and Places

landlocked completely surrounded by land with no direct access to the ocean

Pamirs some of Central Asia's high mountains

Fergana Valley large fertile valley in the plains region of Central Asia

Kara-Kum desert in Turkmenistan

Kyzyl Kum desert in Uzbekistan and Kazakhstan

Aral Sea sea that is actually a large lake, which is shrinking due to irrigation

Section Summary

PHYSICAL FEATURES

Central Asia, the middle part of the continent, is **landlocked**. In the region's east, there are rugged, high mountains. Large glaciers are common in the high mountains. One area of high mountains is called the **Pamirs**.

Because it is landlocked and has such rugged land, Central Asia is isolated. Communication and travel are difficult. The area also has many earthquakes.

From the mountains, the land slowly slopes down to the Caspian Sea in the west. Some land there is 95 feet (29 m) below sea level. The land between the sea and mountains is plains and plateaus. The fertile **Fergana Valley** is in the plains.

Central Asia also has some rivers and lakes. Two important rivers are the Syr Darya (sir durh-YAH) and the Amu Darya (uh-MOO duhr-YAH).

> What two factors make Central Asia isolated?
>
> _______________________
>
> _______________________

> Underline the sentence that names two important rivers in Central Asia.

Interactive Reader and Study Guide

Section 1, *continued*

They make the Fergana Valley fertile. The rivers flow into the Aral Sea, which is really a large lake. Lake Balkhash is also an important lake. It has freshwater at one end and salty water at the other.

Circle the name of the sea that is really a large lake.

CLIMATE AND VEGETATION

Most of Central Asia has a harsh, dry climate. Temperatures range from very cold to very hot, and there is not much rain. It is hard for plants to grow.

The mountain peaks are cold, dry, and windy. There are harsh desert areas between the mountains and sea. Two major deserts are the **Kara-Kum** and **Kyzyl Kum**. The deserts do have some sources of water. Some areas have rivers crossing them, which lets people live there. This lets people irrigate, or supply water to the land.

Underline the names of two major deserts in Central Asia.

Only the far north of Central Asia has a milder climate. Grasses and trees are able to grow there.

NATURAL RESOURCES

Some of Central Asia's natural resources are water, oil, and gas. There is also a supply of minerals, such as gold, lead, and copper.

List some of Central Asia's natural resources:

____

____

____

People use the Syr Darya and Amu Darya rivers to irrigate and make electricity. But water is limited. This has led to conflicts over how to use it. Also, irrigation has kept the rivers from flowing into the **Aral Sea**. As a result, the sea has lost much of its water.

Oil and gas can only help the region if the countries can sell it. There are no ocean ports to transport it, so they need to build and maintain pipelines. But this is hard because of the rugged land, as well as economic and political problems.

CHALLENGE ACTIVITY

Critical Thinking: Analyzing Information Write a fact sheet called *Central Asia: Tips for Travelers*. Include key facts that travelers to the region should know and a list of supplies they should bring.

     Interactive Reader and Study Guide

Central Asia

Section 2

MAIN IDEAS

1. Throughout history, many different groups have conquered Central Asia.
2. Many different ethnic groups and their traditions influence culture in Central Asia.

Key Terms and Places

Samarqand city along the Silk Road that grew rich from trade

nomads people who move often from place to place

yurt moveable round house made of wool felt mats hung over a wood frame

Section Summary

HISTORY

For hundreds of years, many groups of people came through Central Asia. They left lasting influences.

At one time, there were two trade routes through Central Asia. One route went between Europe and India, through Afghanistan. The other route went through the rest of Central Asia. It was called the Silk Road, because traders from Europe traveled it to get silk and spices from China. **Samarqand** and other cities on the Silk Road grew rich.

By 1500, Europeans stopped using these roads. They discovered they could sail to East Asia on the Indian Ocean. The region became isolated and poor.

The Silk Road brought many people into Central Asia. In AD 500, Turkic-speaking nomads came from northern Asia. From the 700s to 1200s, Arabs ruled. They brought their religion, Islam. Then the Mongols conquered Central Asia. After the Mongols, groups such as the Uzbeks, Kazaks, and Turkmen came in.

In the mid-1800s, Russia conquered this region. Russians built railroads. They also increased oil and cotton production. But people began to resist Russia's rule. After the Soviets took power in Russia, they wanted to weaken resistance to their rule. So

> What places did the two trade routes in Central Asia connect?
>
> _______________________
>
> _______________________

> Underline the reason that Europeans stopped using the trade routes.

> Circle the dates that the Arabs ruled Central Asia.

they divided Central Asia into republics. They also encouraged ethnic Russians to move in. The Soviets also built huge irrigation projects for more cotton production. In 1991, the Soviet government collapsed. Central Asia's republics became independent countries.

> **What happened to Asia's republics when the Soviet government collapsed in 1991?**
>
> ___________________________
>
> ___________________________

CULTURE

The people who came through Central Asia brought new languages, religions, and ways of life. These mixed with traditional ways.

For centuries, Central Asians raised herds of horses, cattle, goats, and sheep. Many lived as **nomads.** They moved their herds to different pastures in summer and winter. They also moved their houses. The Central Asian nomad's moveable house is called a **yurt**. It is an important symbol today. Even people in cities put up yurts for special events. Nomads are still common in Kyrgyzstan.

> **Unscramble these letters to identify a feature of nomad life: *tury*. Write your answer:**
>
> ___________________________

Today, most of the region's ethnic groups are part of the larger Turkic group. There are ethnic Russians, also. Each group speaks its own language. Some countries have many languages. In some countries Russian is still the official language, because of earlier Russian rule. The Russians also brought Cyrillic, their alphabet. Now most countries use the Latin alphabet, the one for writing English.

> **Circle the name of the Russian alphabet.**

The region's main religion is Islam, but there are also others. Some people are Russian Orthodox, a Christian religion. Today, many religious buildings that were closed by the Soviets have opened again.

CHALLENGE ACTIVITY

Critical Thinking: Drawing Conclusions What would Central Asia be like today if the Silk Road was still in use? Write a travel journal entry for a trip along the Silk Road describing who and what you saw.

Central Asia

Section 3

MAIN IDEAS

1. The countries of Central Asia are working to develop their economies and to improve political stability in the region.
2. The countries of Central Asia face issues and challenges related to the environment, the economy, and politics.

Key Terms and Places

Taliban radical Muslim group that arose in Afghanistan in the mid-1990s

Kabul capital of Afghanistan

dryland farming farming that relies on rainfall instead of irrigation

arable suitable for growing crops

Section Summary

CENTRAL ASIA TODAY

Central Asia is working to recover from a history of invasions and foreign rulers. The region is trying to build more stable governments and stronger economies.

During the 1980s, Afghanistan was at war with the Soviet Union. In the mid-1990s, the **Taliban** took power. This was a radical Muslim group. It ruled most of the country, including **Kabul**, the capital. It based its laws on strict Islamic teachings. Most people disagreed with the Taliban. A terrorist group based in Afghanistan attacked the United States on September 11, 2001. As a result, U.S. and British forces toppled the Taliban government. Now people in Afghanistan have a constitution and more freedom. But some groups still threaten violence.

Kazakhstan was the first area in Central Asia that Russia conquered. It still has many Russian influences. Its economy suffered when the Soviet Union fell. But it is growing again, because of oil reserves and a free market. Kazakhstan has a stable democratic government. People elect a president and parliament.

> What group ruled Afghanistan from the mid-1990s to 2001?
>
> ___________________

> Underline the sentence that explains why Kazakhstan's economy is growing.

     Interactive Reader and Study Guide

Section 3, *continued*

In Kyrgystan, many people farm. They irrigate or use **dryland farming**. This does not bring much money, but tourism may help the economy. In recent years, there have been government protests.

Tajikistan now has a more stable government, after ending conflicts between different groups. Today, the economy relies on cotton farming. But only about 5 or 6 percent of the land is **arable**.

Turkmenistan's president is elected for life and has all the power. The economy is based on oil, gas, and cotton farming. The country is a desert, but it has the longest irrigation channel in the world.

Uzbekistan's president is also elected and has all the power. The economy is based on oil and cotton. The economy is stable, but not really growing.

> Why should Tajikistan look for other ways to support its economy?
>
> _______________________________
> _______________________________

ISSUES AND CHALLENGES

Central Asia faces challenges in three areas today: environment, economy, and politics.

The shrinking Aral Sea is a serious problem for Central Asia's environment. The seafloor is dry. Dust, salt, and pesticides blow out of it. Its environment also has leftover radiation from Soviet nuclear testing. People's health is a concern. Crop chemicals are also a problem for the environment, harming farmlands.

> Underline the main challenges that Central Asia's environment faces.

Central Asia's economy relies on cotton. This has hurt many of their economies. Oil and gas reserves may bring in more money one day. Today there are still challenges, such as old equipment, that need to be overcome for Central Asia's economy to grow.

Central Asia does not have widespread political stability. In some countries, people do not agree on the best kind of government. Some turn to violence or terrorism as a result.

CHALLENGE ACTIVITY

Critical Thinking: Analyzing Information Imagine you work for an organization that is helping Central Asia clean up its environment. Create a poster that will convince people to donate money to this cause.

   Interactive Reader and Study Guide

History of Ancient Egypt

CHAPTER SUMMARY

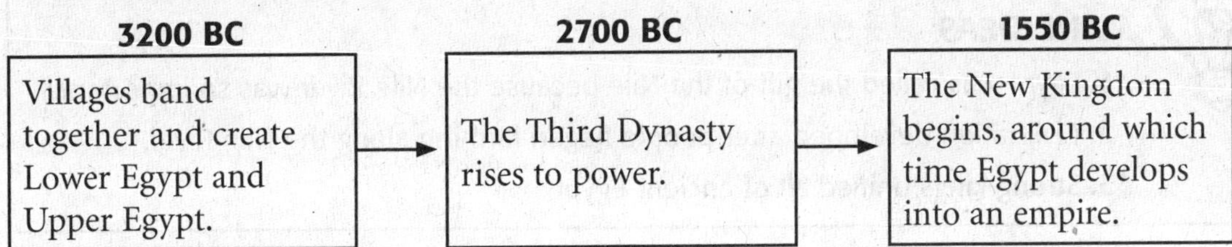

COMPREHENSION AND CRITICAL THINKING

Use information from the graphic organizer to help answer the following questions.

1. Sequence Approximately when did the Third Dynasty rise to power?

____

____

2. Draw Conclusions What was the importance of the rise of the Third Dynasty?

____

____

3. Evaluate What was Egypt like after its reunification around 2050 BC?

____

____

4. Explain How did Egypt become an empire?

____

____

History of Ancient Egypt

Section 1

MAIN IDEAS

1. Egypt was called the gift of the Nile because the Nile River was so important.
2. Civilization developed after people began farming along the Nile River.
3. Strong kings unified all of ancient Egypt.

Key Terms and Places

Nile River important river in Egypt

Upper Egypt southern part of Egypt

Lower Egypt northern part of Egypt

cataracts river rapids

delta triangle-shaped area of land made from soil deposited by a river

pharaoh ruler of ancient Egypt, literally means "great house"

dynasty series of rulers from the same family

Section Summary

THE GIFT OF THE NILE

The existence of Egypt was based solely around the **Nile River**, the world's longest river. The Nile carries water from central Africa through a vast stretch of desert land. The river was so important to people that Egypt was called the gift of the Nile.

Ancient Egypt developed along a 750-mile stretch of the Nile, and was originally organized into two kingdoms—**Upper Egypt** and **Lower Egypt**. Upper Egypt was located upriver in relation to the Nile's flow. Lower Egypt was the northern region and was located downriver.

Cataracts, or rapids, marked the southern border of Upper Egypt. Lower Egypt was centered in the river **delta**, a triangle-shaped area of land made of soil deposited by the river. In midsummer, the Nile would flood Upper Egypt and in the fall the river would flood Lower Egypt. This made sure that the farmland would stay moist and fertile. As the land surrounding the Nile Valley was arid desert,

> **Why is a river a gift to a desert land?**
>
> ___________________________
>
> ___________________________
>
> ___________________________

  Interactive Reader and Study Guide

Section 1, *continued*

this watered area was the lifeline for everyone who lived in the region.

CIVILIZATION DEVELOPS IN EGYPT

With dry desert all around, it is no wonder that ancient settlers were attracted to this abundant and protected area of fertile farmland. Hunter-gatherers first moved to the area around 12,000 years ago and found plenty of meat and fish to hunt and eat. By 4500 BC farmers were living in villages and growing wheat and barley. They were also raising cattle and sheep.

Around 3200 BC the Egyptian villages became organized into two kingdoms. The capital of Lower Egypt was located in the northwest Nile Delta at a town called Pe. The capital city of Upper Egypt was called Nekhen. It was located on the west bank of the Nile.

KINGS UNIFY EGYPT

Around 3100 BC Menes (MEE-neez), the king of Upper Egypt, invaded Lower Egypt. He married a princess there in order to unite the two kingdoms under his rule. Menes was the first **pharaoh**, which literally means ruler of a "great house." He also started the first Egyptian **dynasty**, or series of rulers from the same family. He built a new capital city, Memphis, which became a popular cultural center. His dynasty ruled for nearly 200 years.

> **Why do you think Menes wanted to unite the two kingdoms?**
>
> ______________________________
>
> ______________________________
>
> ______________________________
>
> ______________________________

CHALLENGE ACTIVITY

Critical Thinking: Drawing Inferences Villages did not develop until people could stop being hunter-gatherers and start growing their own food. From villages came powerful leaders who united larger territories and people under one organization. Imagine that you are an ancient Egyptian interested in becoming a leader. Write a speech explaining what would make you a powerful person fit for ruling a large village.

     Interactive Reader and Study Guide

Section 2

MAIN IDEAS

1. Life in the Old Kingdom was influenced by pharaohs, roles in society, and trade.
2. Religion shaped Egyptian life.
3. The pyramids were built as tombs for Egypt's pharaohs.

Key Terms and Places

Old Kingdom a period in Egyptian history that lasted from about 2700 to 2200 BC

nobles people from rich and powerful families

afterlife life after death, a widely held ancient Egyptian belief

mummies specially treated bodies wrapped in cloth

elite people of wealth and power

pyramids huge, stone tombs with four triangle-shaped walls that meet at a top point

engineering application of scientific knowledge for practical purposes

Section Summary

LIFE IN THE OLD KINGDOM

Around 2700 BC the Third Dynasty began a period in Egyptian history known as the **Old Kingdom**. During the next 500 years, the Egyptians developed a political system based on the belief that the pharaoh was both a king and a god. The most famous pharaoh of the Old Kingdom was Khufu, in whose honor the largest of the pyramids was built.

Although the pharaoh owned everything, he was also held personally responsible if anything went wrong. He was expected to make trade profitable and prevent war. To manage these duties, he appointed government officials, mostly from his family. Social classes developed, with the pharaoh at the top and **nobles** from rich and powerful families making up the upper class. The middle class included some government officials, scribes, and rich craftspeople. Most people, including farmers, belonged to the lower class. Lower-class people were often used by the pharaoh as labor.

> Would you say that there was any distinction between religion and politics in Egypt's Old Kingdom? Why or why not?
>
> _______________________
> _______________________
> _______________________

> Of the upper, middle, and lower classes, which was the largest in ancient Egypt?
>
> _______________________
> _______________________

Section 2, continued

Trade also developed during the Old Kingdom. Traders sailed on the Mediterranean and south on the Nile and the Red Sea to acquire gold, copper, ivory, slaves, wood, and stone.

RELIGION AND EGYPTIAN LIFE

The Old Kingdom formalized a religious structure that everyone was expected to follow. Over time, certain cities built temples and were associated with particular gods.

Much of Egyptian religion focused on the **afterlife**. Each person's *ka* (KAH), or life force, existed after death, but remained linked to the body. To keep the *ka* from suffering, the Egyptians developed a method called embalming to preserve bodies. Royalty had their bodies preserved as **mummies**, specially treated bodies wrapped in cloth. Other members of the **elite** also had their bodies preserved.

THE PYRAMIDS

Pyramids, spectacular stone monuments, were built to house dead rulers. Many pyramids are still standing today, amazing reminders of Egyptian **engineering**.

CHALLENGE ACTIVITY

Critical Thinking: Drawing Inferences Think about the way in which Egyptians viewed the pharaoh. Then think about how we view our current U.S. President. In what ways are these views similar? In what ways are they different? Write a one-page essay considering whether a god-king pharaoh ruling today would be loved or hated by his people.

Interactive Reader and Study Guide

Section 3

MAIN IDEAS

1. The Middle Kingdom was a period of stable government between periods of disorder.

2. The New Kingdom was the peak of Egyptian trade and military power, but its greatness did not last.

3. Work and daily life differed among Egypt's social classes.

Key Terms and Places

Middle Kingdom period of stability and order in ancient Egypt between about 2050 and 1750 BC

New Kingdom the height of Egypt's power and glory, between 1550 and 1050 BC

Kush kingdom south of Egypt

trade routes paths followed by traders

Section Summary

THE MIDDLE KINGDOM

The Old Kingdom ended with the pharaohs in debt. Ambitious nobles serving in government positions managed to take power from the pharaohs and rule Egypt for nearly 160 years. Finally, a powerful pharaoh regained control of Egypt around 2050 BC and started a peaceful period of rule. This era was called the **Middle Kingdom** and lasted until Southwest Asian invaders conquered Lower Egypt around 1750 BC.

> From where did the raiders who ended the Middle Kingdom come?
> ___________________________
> ___________________________

THE NEW KINGDOM

When an Egyptian named Ahmose (AHM-ohs) drove away the invaders and declared himself king of Egypt in 1550 BC, he ushered in Egypt's eighteenth dynasty and the start of the **New Kingdom**. Responding to invasions, Egypt took control of possible invasion routes by taking over areas such as Syria and **Kush**, and quickly became the leading military power in the region, with an empire extending from the Euphrates River in the northeast to Nubia in

Section 3, *continued*

the south. These conquests also made Egypt rich, through gifts and vastly expanded **trade routes**. One ruler in particular, Queen Hatshepsut, was active in establishing new paths for traders.

Despite the strong leadership of Ramses the Great, a tide of invasions from Southwest Asia and from the west eventually reduced Egypt to violence and disorder.

> Which direction would you go from Egypt to reach Nubia?
> _______________________
> _______________________

WORK AND DAILY LIFE

During the Middle and New Kingdoms, Egypt's population continued to grow and become more complex. Professional and skilled workers like scribes, artisans, artists, and architects were honored. These roles in society were usually passed on in families, with young boys learning a trade from their fathers.

> Which type of workers likely designed the pyramids?
> _______________________
> _______________________

For farmers and peasants, who made up the vast majority of the population, life never changed. In addition to hard work on the land, they were required to pay taxes and were subject to special labor duty at any time. Only slaves were beneath them in social status.

> For farmers, did daily life in Egypt change much with the rise and fall of dynasties and kingdoms?
> _______________________
> _______________________

Most Egyptian families lived in their own homes. Boys were expected to marry young and start their own families. Women focused on the home, but many also had jobs outside the home. Egyptian women had the legal rights to own property, make contracts, and divorce their husbands.

CHALLENGE ACTIVITY

Critical Thinking: Drawing Inferences Design a "want ad" for a position held in ancient Egyptian society. Then write a letter to a potential employer explaining why you should be hired.

History of Ancient Egypt

Section 4

MAIN IDEAS

1. Egyptian writing used symbols called hieroglyphics.
2. Egypt's great temples were lavishly decorated.
3. Egyptian art filled tombs.

Key Terms

hieroglyphics Egyptian writing system, one of the world's first, which used symbols

papyrus long-lasting, paper-like substance made from reeds

Rosetta Stone a stone slab discovered in 1799, that was inscribed with hieroglyphics and their Greek meanings

sphinxes imaginary creatures with the bodies of lions and the heads of other animals or humans

obelisk a tall, four-sided pillar that is pointed on top

Section Summary

EGYPTIAN WRITING

Egyptians invented one of the world's first writing systems, using a series of images, symbols, and pictures called **hieroglyphics** (hy-ruh-GLIH-fiks). Each symbol represented one or more sounds in the Egyptian language.

At first hieroglyphics were carved in stone. Later, they were written with brushes and ink on **papyrus** (puh-PY-ruhs). Because papyrus didn't decay, many ancient Egyptian texts still survive, including government records, historical records, science texts, medical manuals, and literary works such as *The Book of the Dead*. The discovery of the **Rosetta Stone** in 1799 provided the key to reading Egyptian writing, as its text was inscribed both in hieroglyphics and in Greek.

> **What language helped scholars to understand the meaning of hieroglyphics on the Rosetta Stone?**
>
> ____________________________
>
> ____________________________

EGYPT'S GREAT TEMPLES

Egyptian architects are known not only for the pyramids but also for their magnificent temples. The temples were lavishly designed with numerous

Section 4, *continued*

statues and beautifully painted walls and pillars. **Sphinxes** and **obelisks** were usually found near the entrances to the temples.

EGYPTIAN ART

Ancient Egyptians were masterful artists and many of their greatest works are found in either the temples or the tombs of the pharaohs. Most Egyptians, however, never saw these paintings, because only kings, priests, or other important people could enter these places.

Egyptian paintings depict a variety of subjects, from crowning kings to illustrating religious rituals to showing scenes from daily life. The paintings also have a particular style, with people drawn as if they were twisting as they walked, and in different sizes depending upon their stature in society. In contrast, animals appear more realistically. The Egyptians were also skilled stone and metal workers, creating beautiful statues and jewelry.

Much of what we know about Egyptian art and burial practices comes from the tomb of King Tutankhamen, one of the few Egyptian tombs that was left untouched by raiders looking for valuables. The tomb was discovered in 1922.

> **Who got to see ancient Egyptian sculpture and painting?**
> ___________________________
> ___________________________
> ___________________________

> **Why is King Tutankhamen's tomb so important for the study of Egyptian history?**
> ___________________________
> ___________________________
> ___________________________
> ___________________________

CHALLENGE ACTIVITY

Critical Thinking: Drawing Inferences Using the library or an online resource, find a key to translate Egyptian hieroglyphics into English. Write a message using hieroglyphics and trade with another student to see if you can read each other's messages. Be sure to provide a copy of your message and the translation to your teacher.

 Interactive Reader and Study Guide

History of Ancient Kush

CHAPTER SUMMARY

Egypt	Kush
women worked in the home	women worked in ________________
led by pharaohs (male)	led by ________________
worshipped ________________, with a human head and man's body	worshipped lion-headed god ________________
developed pictograph writing style called ________________	developed pictograph writing style called Meroitic
built ________________ pyramids to bury dead kings	built ________________ pyramids to bury dead kings

COMPREHENSION AND CRITICAL THINKING

Use the answers to the following questions to fill in the graphic organizer above.

1. Explain What were three similarities between the Kush and Egyptian cultures?

____

____

2. Identify Cause and Effect How was the position of women in Kush society different than that of Egyptian women?

____

____

3. Evaluate Why do you think people's houses were different in Kush and Egypt?

____

____

4. Draw a Conclusion How are the two cultures alike? How are they different?

____

____

History of Ancient Kush

Section 1

MAIN IDEAS

1. Geography helped early Kush civilization develop in Nubia.
2. Egypt controlled Kush for about 450 years.
3. After winning its independence, Kush ruled Egypt and set up a new dynasty there.

Key Terms and Places

Nubia a region in northeast Africa where the kingdom of Kush developed

ebony a type of dark, heavy wood

ivory a white material taken from elephant tusks

Section Summary

GEOGRAPHY AND EARLY KUSH

The kingdom of Kush developed south of Egypt along the Nile, in the region we now call **Nubia**. Every year, floods provided a rich layer of fertile soil. Farming villages thrived. The area was also rich in minerals such as gold, copper, and stone. These resources contributed to the region's wealth.

> **What valuable minerals were important to Kush's prosperity?**
>
> ___________________
>
> ___________________
>
> ___________________

Over time some rich farmers became leaders of their villages. Around 2000 BC, one of these leaders took control of other villages and made himself king of Kush.

> **Around what year did the first king of Kush appear?**
>
> ___________________
>
> ___________________

The kings of Kush ruled from their capital at Kerma (KAR-muh). The city was located on the Nile just south of a cataract, or stretch of rapids. Because the Nile's cataracts made parts of the river hard to pass through, they were natural barriers against invaders.

As time passed Kushite society became more complex. In addition to farmers and herders, some people of Kush became priests or artisans.

Interactive Reader and Study Guide

EGYPT CONTROLS KUSH

Kush and Egypt were neighbors and trading partners. The Kushites sent slaves to Egypt. They also sent gold, copper, and stone, as well as the prized materials **ebony** and **ivory**.

Relations between Kush and Egypt were not always peaceful, however. Around 1500 BC Egyptian armies under the pharaoh Thutmose I invaded and conquered most of Nubia, including all of Kush. The Kushite palace at Kerma was destroyed. Kush remained an Egyptian territory until the mid-1000s BC, when the Kushite leaders regained control.

> For about how many years was Kush under Egyptian control?
>
> _______________________
>
> _______________________

KUSH RULES EGYPT

By around 850 BC, Kush was once again as strong as it had been before it was conquered by Egypt. During the 700s, under the king Kashta, the Kushites began to invade Egypt. Kashta's son, Piankhi (PYANG-kee), believed that the gods wanted him to rule all of Egypt. By the time he died in 716 BC, Piankhi had accomplished this task. His kingdom extended north from Napata all the way to the Nile Delta.

Piankhi's brother, Shabaka (SHAB-uh-kuh), declared himself pharaoh and began the Twenty-fifth, or Kushite, Dynasty in Egypt. Egyptian culture thrived during the Twenty-fifth Dynasty. In the 670s BC, however, the powerful army of the Assyrians from Mesopotamia invaded Egypt. The Assyrians' iron weapons were better than the Kushites' bronze weapons. The Kushites were slowly pushed out of Egypt.

> What metals did the Assyrians and Kushites use to make weapons?
>
> _______________________
>
> _______________________

CHALLENGE ACTIVITY

Critical Thinking: Making Judgments Some leaders do not take control of other lands and people, even though they have the power to do so. What does this tell you about village leaders who make themselves kings over whole regions?

History of Ancient Kush

Section 2

MAIN IDEAS

1. Kush's economy grew because of its iron industry and trade network.
2. Some elements of Kushite society and culture were borrowed from other cultures while others were unique to Kush.
3. The decline and defeat of Kush was caused by both internal and external factors.

Key Terms and Places

Meroë economic center of Kush, new Kushite capital

trade network a system of people in different lands who trade goods back and forth

merchants traders

exports items sent to other regions for trade

imports goods brought in from other regions

Section Summary

KUSH'S ECONOMY GROWS

After they lost control of Egypt, the people of Kush devoted themselves to increasing agriculture and trade, hoping to make their country rich again. The economic center of Kush during this period was **Meroë** (MER-oh-wee). Gold could be found nearby, as could forests of ebony and other wood. In this rich location the Kushites developed Africa's first iron industry. Iron ore and wood for furnaces were easily available, so the iron industry grew quickly.

> **What industry helped make Kush a rich and successful kingdom again?**
>
> _______________________
> _______________________

In time, Meroë became the center of a large **trade network**. The Kushites sent goods down the Nile to Egypt. From there, Egyptian and Greek **merchants** carried goods to ports on the Mediterranean and Red seas, and to southern Africa. These goods may have eventually reached India and perhaps China. Kush's **exports** included gold, pottery, iron tools, ivory, leopard skins, ostrich feathers, elephants, and slaves. **Imports** included fine jewelry and luxury items from Egypt, Asia, and lands along the Mediterranean.

> **What direction is "down the Nile?"**
>
> _______________________
> _______________________

Section 2, *continued*

SOCIETY AND CULTURE

The most obvious influence on Kush during this
period was Egyptian, but many elements of Kushite
culture were not borrowed from anywhere else.
The people of Kush worshipped their own gods and
even developed their own written language. Women
were expected to be active in their society. Some
women rose to positions of great authority, especially
in religion. A few women, such as Queen
Shanakhdakheto (shah-nahk-dah-KEE-toh), even
ruled the empire alone.

> How was the position of women
> in Kushite society different than
> that of women in most other
> ancient civilizations?
>
> _______________________
>
> _______________________
>
> _______________________

DECLINE AND DEFEAT

Kushite civilization centered at Meroë reached its
height in the first century BC. Eventually it fell due
to both external and internal factors. The stores of
iron and other metals dwindled, and the overgrazing
of cattle caused a deterioration of farmland. Another
powerful trading center, Aksum (AHK-soom), located
in modern-day Ethiopia and Eritrea, began competing
with Kush. Soon trade routes were bypassing Meroë
for Aksum. After Aksum had decimated Kush
economically, the Aksumite leader King Ezana
(AY-zah-nah) sent an invading army and conquered
the once-powerful Kush.

> Circle the name and kingdom of
> the ruler who eventually defeated
> Kush.

CHALLENGE ACTIVITY

Critical Thinking: Drawing Inferences You are a Kushite leader in the
600s BC. Write a short essay explaining your plan to make the kingdom
of Kush rich and powerful again.

History of West Africa

CHAPTER SUMMARY

West African Timeline

c. 300 AD	Small bands of _______________ defend themselves against _______________.
c. 800 AD	Ghana controls _______________.
c. 1200 AD	Ghana's decline is caused by _______________, _______________, and _______________.
c. 1300 AD	In Mali, Mansa Musa establishes Timbuktu as a great center of Islamic culture.
c. 1400s AD	_______________ empire declines; _______________ empire grows powerful under Sunni Ali.
c. 1600 AD	_______________ destroy Timbuktu and Gao; era of West African empires ends.

COMPREHENSION AND CRITICAL THINKING

Use the answers to the following questions to complete the sentences in
the graphic organizer above.

1. Evaluate Why did West African society change dramatically in 300 AD?

____

____

2. Identify Cause and Effect How did Ghana become the first West African empire?
What were the causes of Ghana's decline?

____

____

3. Make Inferences In the 1400s, why did one empire gain power as another declined?

____

____

4. Identify Cause and Effect What led to the fall of the Songhai empire?

____

____

Section 1

MAIN IDEAS

1. Ghana controlled trade and became wealthy.
2. Through its control of trade, Ghana built an empire.
3. Attacking invaders, overgrazing, and the loss of trade caused Ghana's decline.

Key Terms

silent barter a process in which people exchange goods without ever contacting each other directly

Section Summary

GHANA CONTROLS TRADE

The empire of Ghana (GAH-nuh) became powerful by controlling Saharan trade routes. Ghana lay between the Niger and Senegal rivers in sub-Saharan Africa, northwest of the nation now called Ghana.

Historians think the first people in Ghana were farmers. Starting around 300, these farmers were threatened by nomadic herders. The herders wanted the water and pastures. For protection, small groups began to band together. These groups grew stronger with the introduction of farming tools and weapons made of iron.

Ghana's territory lay between the desert and the forests. These were areas rich with salt and gold. The gold and salt trade sometimes followed a process called **silent barter**. In this process people exchange goods without contacting each other directly. This ensured peaceful business and kept the location of the gold mines secret.

As populations grew and trade increased, the rulers of Ghana grew stronger. Their armies used iron weapons. They took control of the trade routes that had been run by North African merchants.

What helped Ghana become a powerful empire?

____

____

Which was more valuable, salt or gold? Why?

____

____

____

____

GHANA BUILDS AN EMPIRE

By 800, Ghana was firmly in control of West Africa's trade routes. As a result, trade became safer and Ghana's influence increased. Traders were charged a tax to enter or leave Ghana. The kings made it illegal for anyone other than themselves to own gold. They also taxed the people of Ghana.

The kings increased the size of Ghana by conquering other tribes. However, Ghana's kings allowed former rulers to keep much of their own power. These kings acted as governors of their territories. The empire of Ghana reached its peak under Tunka Manin (TOOHN-kah MAH-nin).

> When the kings made it illegal for anyone else to own gold, what happened to the value of gold? Explain.
>
> ___________________________
> ___________________________
> ___________________________

GHANA'S DECLINE

By the end of the 1200s, Ghana had collapsed. Three major factors contributed to its decline. A group of Muslim Berbers called the Almoravids invaded and weakened the empire. These Berbers were herders. Their animals overgrazed and ruined the farmland. Many farmers left. At the same time, internal rebellions led to loss of order in Ghana.

> List two reasons for the decline of Ghana's empire.
>
> ___________________________
> ___________________________
> ___________________________

CHALLENGE ACTIVITY

Critical Thinking: Drawing Inferences Recreate the silent barter system in the classroom. Divide students into groups of gold and salt traders. Each group of "traders" should write a one-page paper detailing the advantages and disadvantages of silent barter.

Interactive Reader and Study Guide

History of West Africa

Section 2

MAIN IDEAS

1. The empire of Mali reached its height under the ruler Mansa Musa, but the empire fell to invaders in the 1400s.

2. The Songhai built a new Islamic empire in West Africa, conquering many of the lands that were once part of Mali.

Key Terms and Places

Niger River river that went through Ghana and Mali

Timbuktu important trade city in Mali

mosque building for Muslim prayer

Gao capital of Songhai

Djenné center of learning in Songhai

Section Summary

MALI

Like Ghana, Mali (MAH-lee) lay along the upper **Niger River**. Mali's location on the Niger River allowed its people to control trade on the river. Mali's rise to power began under a ruler named Sundiata (soohn-JAHT-ah).

A cruel ruler conquered Mali when Sundiata was a boy. When Sundiata grew older, he raised an army and won Mali's independence. Sundiata conquered nearby kingdoms, including Ghana, and took over the salt and gold trades. He also took over religious and political authority held by local leaders.

Mali's greatest and most famous ruler was a Muslim named Mansa Musa (MAHN-sah moo-SAH). Under his leadership, Mali reached its peak. Musa ruled Mali for about 25 years and captured many important trading cities, including **Timbuktu**. He also made the Islamic world aware of Mali on his pilgrimage to Mecca.

Mansa Musa stressed the importance of learning Arabic in order to read the Qur'an. He spread Islam through West Africa by building **mosques** in cities.

> **What river flowed through both Ghana and Mali?**
> _______________________
> _______________________

> **Name three important things Mansa Musa did as leader of Mali.**
> _______________________
> _______________________
> _______________________
> _______________________

After Mansa Musa died, invaders destroyed the schools and mosques of Timbuktu. Nomads from the Sahara seized the city. By 1500 nearly all of the lands the empire had once ruled were lost.

SONGHAI

As Mali was reaching its height, the neighboring Songhai (SAHNG-hy) kingdom was also growing in strength. In the 1300s, the Songhai lands, including **Gao**, its capital, lay within the empire of Mali. As Mali weakened, the Songhai broke free. Songhai leader Sunni Ali (SOOH-nee ah-LEE) strengthened and enlarged the Songhai empire.

After Sunni Ali died, his son Sunni Baru became ruler. He was not Muslim. The Songhai people feared that if Sunni Baru did not support Islam they would lose trade, so they rebelled. After overthrowing Sunni Baru, the leader of that rebellion became known as Askia the Great.

Muslim culture and education thrived during Askia's reign. Timbuktu's universities, schools, libraries, and mosques attracted thousands. **Djenné** was also an important center of learning.

Morocco invaded Songhai and destroyed Gao and Timbuktu. Songhai never recovered and trade declined. Other trade centers north and south of the old empire became more important. The period of great West African empires came to an end.

CHALLENGE ACTIVITY

Critical Thinking: Drawing Inferences You are a reporter who does not know much about Africa. One day, the ruler of Mali or Songhai comes through your city. Write an article about this person.

> **Why was Sunni Baru overthrown?**
> ___________________
> ___________________
> ___________________
> ___________________

> **Do research on the Internet or in a library to find the population of Timbuktu today. Write that figure here:**
> ___________________
> ___________________

History of West Africa

Section 3

MAIN IDEAS

1. West Africans have preserved their history through storytelling and the written accounts of visitors.

2. Through art, music, and dance, West Africans have expressed their creativity and kept alive their cultural traditions.

Key Terms

oral history a spoken record of past events

griots West African storytellers responsible for reciting oral history

proverbs short sayings of wisdom or truth

kente handwoven, brightly colored cloth made in West Africa

Section Summary

PRESERVING HISTORY

Writing was not common in West Africa. None of the major early civilizations of West Africa developed a written language. Arabic was the only written language used. Instead of writing their history, West Africans passed along information about their civilization through **oral history** in their native languages.

> Did Arabic replace the native languages of the West Africans? How do you know your answer is correct?
>
> _______________________
> _______________________
> _______________________
> _______________________

The task of remembering and telling West Africa's history was entrusted to storytellers called **griots** (GREE-ohz). Griots tried to make their stories entertaining. They also told **proverbs**, or short sayings of wisdom or truth. The griots had to memorize hundreds of names and dates. Some griots confused names and events in their heads, so some facts became distorted. Still, much knowledge could be gained by listening to a griot.

> Why might the history of the griots not be perfectly accurate?
>
> _______________________
> _______________________
> _______________________
> _______________________

Some griot poems are epics, long poems about kingdoms and heroes. Many of these poems were collected in the *Dausi* (DAW-zee) and the *Sundiata*. The *Dausi* tells the history of Ghana, but it also includes myths and legends. The *Sundiata* tells the story of Mali's great ruler. A conqueror killed his

     Interactive Reader and Study Guide

Section 3, *continued*

family, but the boy was spared because he was sick. He grew up to be a great warrior and overthrew the conqueror.

Though the West Africans left no written histories, visitors from other parts of the world did write about the region. Much of what we know about early West Africa comes from the writings of travelers and scholars from Muslim lands such as Spain and Arabia. Ibn Battutah was the most famous visitor to write about West Africa.

ART, MUSIC, AND DANCE

Besides storytelling, West African cultures considered other art forms, including sculpture, mask-making, cloth-making, music, and dance just as important. West African artists made sculptures of people from wood, brass, clay, ivory, stone, and other materials. Some of these images have inspired modern artists like Henri Matisse and Pablo Picasso.

> Circle the names of the modern artists inspired by the images crafted by West African sculptors.

West Africans are also known for distinctive mask-making and textiles. Particularly prized is the brightly colored **kente** (ken-TAY), a hand-woven cloth that was worn by kings and queens on special occasions.

In many West African societies, music and dance were as important as the visual arts. Singing, dancing, and drumming were great entertainment, but they also helped people celebrate their history and were central to many religious celebrations.

> List three ways in which music had a place in West African culture.
>
> _______________________
> _______________________
> _______________________
> _______________________

CHALLENGE ACTIVITY

Critical Thinking: Drawing Inferences Much of what we know about West Africa comes from oral traditions or accounts by visitors to the land. Write a short essay evaluating the accuracy of these resources. Which sources are primary? Which are secondary? Consider how much a visitor who was not raised in a culture can really understand about that culture.

North Africa

CHAPTER SUMMARY

North Africa

Egypt	Countries of the Maghreb
It has large areas of desert.	They have large areas of desert.
The Nile supports fertile farmland.	The coastal areas support agriculture.
Nomadic Bedouins herd livestock.	Nomadic Tuareg herd livestock.
The Suez Canal is a strategic waterway.	The Mediterranean connects these countries to Europe.
Most people are Arabs.	Most people are Arabs.
Most people practice Islam.	Most people practice Islam.
Oil, textiles, and tourism are major industries.	Oil is most important industry. Mining and metal work are important, too.
Cairo and Alexandria are major cities.	Casablanca, Tangier, Algiers, Tunis, and Tripoli are major cities.
Ancient Egyptians built pyramids and created hieroglyphics.	Early Berbers herded livestock and grew wheat and barley.
From the 600s to the 1800s, Arabs from Southwest Asia controlled Egypt.	From the 600s to the 1800s, Arabs from Southwest Asia controlled the Maghreb.
French and British interests controlled it during the 1800s.	Spanish, French, and Italian interests controlled it during the 1800s.
The role of Islam is a challenge today.	The role of Islam is a challenge today.

COMPREHENSION AND CRITICAL THINKING

Use information from the chart to help answer the following questions.

1. Identify Name the capitals of Algeria, Tunisia, Libya, and Egypt.

____

2. Make Inferences Why are oases important to the nomadic Bedouins and Tuareg?

____

3. Contrast How does farming in Egypt differ from farming in the Maghreb?

____

4. Make Judgments/Evaluate What do you think is the most important Arab influence on North African life? Why?

____

North Africa

Section 1

MAIN IDEAS

1. Major physical features of North Africa include the Nile River, the Sahara, and the Atlas Mountains.
2. The climate of North Africa is hot and dry, and water is the region's most important resource.

Key Terms and Places

Sahara world's largest desert, covering most of North Africa

Nile River the world's longest river, located in Egypt

silt finely ground, fertile soil good for growing crops

Suez Canal strategic waterway connecting the Mediterranean and Red Seas

oasis wet, fertile area in a desert where a natural spring or well provides water

Atlas Mountains mountain range on the northwestern side of the Sahara

Section Summary

PHYSICAL FEATURES

Morocco, Algeria, Tunisia, Libya, and Egypt are the five countries of North Africa. All five countries have northern coastlines on the Mediterranean Sea. The largest desert in the world, the **Sahara**, covers most of North Africa.

The **Nile River**, the world's longest, flows northward through the eastern Sahara. Near its end, it becomes a large river delta that empties into the Mediterranean Sea. The river's water irrigates the farmland along its banks. In the past, flooding along the Nile left finely ground fertile soil, called **silt**, in the surrounding fields. Today, the Aswan High Dam controls flooding and prevents silt from being deposited in the nearby fields. As a result, farmers must use fertilizer to aid the growth of crops. East of the Nile River is the Sinai Peninsula, which is made up of rocky mountains and desert. The Sinai is separated from the rest of Egypt by the **Suez Canal**, a strategic waterway that connects the Mediterranean Sea with the Red Sea.

> Name the five countries of North Africa.
> ___________________________
> ___________________________

> Describe the Nile River.
> ___________________________
> ___________________________
> ___________________________

   Interactive Reader and Study Guide

Section 1, *continued*

The Sahara has a huge impact on all of North Africa. It is made up of sand dunes, gravel plains, and rocky, barren mountains. Because of the Sahara's harsh environment, few people live there. Small settlements of farmers are located by **oases**—wet, fertile areas in the desert that are fed by natural springs. The Ahaggar Mountains are located in central North Africa. The **Atlas Mountains** are in the northwestern part of North Africa.

> Why would an oasis be valuable to someone traveling in the desert?
> _______________________
> _______________________

CLIMATE AND RESOURCES

Most of North Africa has a desert climate. It is hot and dry during the day, and cool or cold during the night. There is very little rain. Most of the northern coast west of Egypt has a Mediterranean climate. There it is hot and dry in the summer, and cool and moist in the winter. Areas between the coast and the Sahara have a steppe climate.

> What kind of climate covers most of North Africa?
> _______________________

Important resources include oil and gas, particularly for Libya, Algeria, and Egypt. In Morocco, iron ore and minerals are important. Coal, oil, and natural gas are found in the Sahara.

CHALLENGE ACTIVITY

Critical Thinking: Evaluating Why do you think almost all of Egypt's population lives along the Nile River? Write a brief paragraph that explains your answer.

 Interactive Reader and Study Guide

North Africa

Section 2

MAIN IDEAS

1. North Africa's history includes ancient Egyptian civilization.
2. Islam influences the cultures of North Africa and most people speak Arabic.

Key Terms and Places

Alexandria city in Egypt founded by Alexander the Great in 332 BC
Berbers an ethnic group who are native to North Africa and speak Berber languages

Section Summary

NORTH AFRICA'S HISTORY

Around 3200 BC people along the northern Nile united into a single Egyptian Kingdom. The ancient Egyptians participated in trade, developed a writing system, and built pyramids in which to bury their pharaohs, or kings. The pyramids were made of large blocks of stone that were probably rolled on logs to the Nile and then moved by barge to the building site. The Great Pyramid of Egypt took about twenty years to finish.

> **Where did ancient Egyptians bury their kings?**
> _______________________________

Hieroglyphs, pictures and symbols that stand for ideas and words, formed the basis for Egypt's first writing system. Each symbol stood for one or more sounds in the Egyptian language. Many writings recorded the achievements of pharaohs.

> **What are hieroglyphs?**
> _______________________________
> _______________________________
> _______________________________

Invaders of North Africa included people from the eastern Mediterranean, Greeks, and Romans. Alexander the Great, the Macedonian king, founded the city of **Alexandria** in Egypt in 332 BC. It became an important port of trade and a great center of learning. Arab armies from Southwest Asia started invading North Africa in the AD 600s. They ruled most or all of North Africa until the 1800s, bringing the Arabic language and Islam to the region.

> **How long did Arabs from Southwest Asia rule North Africa?**
> _______________________________

Section 2, *continued*

In the 1800s European countries began invading North Africa. By 1912 Spain and France controlled Morocco, France also controlled Tunisia and Algeria, Italy controlled Libya, and the British controlled Egypt. The countries gradually gained independence in the mid-1900s. Algeria was the last country to win independence in 1962. Today the countries of North Africa are trying to build stronger ties to other Arab countries.

What European countries ruled North Africa in the early 1900s?

____

____

CULTURES OF NORTH AFRICA

Egyptians, **Berbers**, and Bedouins make up almost all of Egypt's population. People west of Egypt are mostly of mixed Arab and Berber ancestry. Most North Africans speak Arabic and are Muslims.

What language do most North Africans speak? What religion do they practice?

____

____

Grains, vegetables, fruits, and nuts are common foods. Couscous, a pellet-like pasta made from wheat, is served steamed with vegetables or meat. Another favorite dish is *fuul*, made from fava beans.

Two important holidays are Muhammad's birthday and Ramadan, a holy month during which Muslims fast. Traditional clothing is long and loose. Many women cover their entire body except for the face and hands.

Name two important North African holidays.

____

____

North Africa is known for its beautiful architecture, wood carving, carpets, and hand-painted tiles. The region has produced important writers, including Egypt's Nobel Prize winner Naguib Mahfouz. Egypt also has a thriving film industry. North African music is based on a scale containing more notes than the one common in Western music, which creates a wailing or wavering sound. The three-stringed sintir of Morocco is a popular instrument.

CHALLENGE ACTIVITY

Critical Thinking: Evaluating Imagine that you are spending your summer vacation traveling throughout North Africa. Write a letter to a friend at home that describes the people you meet and the places you visit.

North Africa

Section 3

MAIN IDEAS

1. Many of Egypt's people are farmers and live along the Nile River.
2. People in the other countries of North Africa are mostly pastoral nomads or farmers, and oil is an important resource in the region.

Key Terms and Places

Cairo capital of Egypt, located in the Nile Delta

Maghreb collective name for Western Libya, Tunisia, Algeria, and Morocco

souks large marketplaces

free port a city in which almost no taxes are placed on goods sold there

dictator someone who rules a country with complete power

Section Summary

EGYPT

More than half of all Egyptians live in rural areas. Most rural Egyptians own small farms or work on large ones owned by powerful families. **Cairo**, Egypt's capital and largest city, is located in the Nile Delta. Overcrowding, limited housing, and pollution are serious problems in Cairo. Alexandria, Egypt's second-largest city, is a major seaport and industrial center. Oil and tourism are important industries in Egypt. Revenue from the Suez Canal provides another source of income. Cotton is an important crop in the Nile Delta. Vegetables, grain, and fruit are grown along the Nile River.

Egypt faces important challenges today. Fertilizing farmland is expensive. Poverty, illiteracy, disease, and pollution are other problems. Still another is the role of Islam. Egyptians disagree on the extent to which Muslim beliefs should influence government. These disagreements have led to violence at times.

> Name three serious problems in Cairo.
>
> _______________________
>
> _______________________
>
> _______________________

> What are some challenges that Egypt faces today?
>
> _______________________
>
> _______________________
>
> _______________________

OTHER COUNTRIES OF NORTH AFRICA

Western Libya, Tunisia, Algeria, and Morocco are known as the **Maghreb**. The Sahara covers most of this region. The major cities and most of the farmland lie along the Mediterranean coast.

Algeria's capital, Algiers, includes an old district called the Casbah. Marketplaces called **souks** jam the narrow streets of this district. In Algeria, as in Egypt, disagreement over the role of Islam in society has led to violence at times.

Tunisia's capital and largest city is Tunis. Tunisia, like other North African countries, has close economic ties to Europe. About two thirds of its imports are from the European Union.

The largest city in Morocco is Casablanca. Tangier, overlooking the Strait of Gibraltar to Spain, is a **free port**. Almost no taxes are charged on goods sold there. Morocco has little oil, but it is an important producer of fertilizer.

More than 85 percent of Libya's population lives in cities. The two largest cities are Benghazi and the capital, Tripoli. The **dictator**, Muammar al-Gadhafi, rules Libya. Because of his support of terrorist activities, Libya's economic relationship with other countries has been hurt.

Oil is the most important industry in North Africa. Mining and tourism are important too. The region's farmers grow and export grains, olives, fruits, and nuts.

> **Why is Europe important to the economies of North Africa?**
> ____________________
> ____________________
> ____________________

> **What is the most important industry in North Africa?**
> ____________________

CHALLENGE ACTIVITY

Critical Thinking: Making Judgments What do you think is the greatest challenge facing the countries of North Africa today? Write a letter to the editor of a newspaper that identifies this challenge and suggests what to do about it.

West Africa

CHAPTER SUMMARY

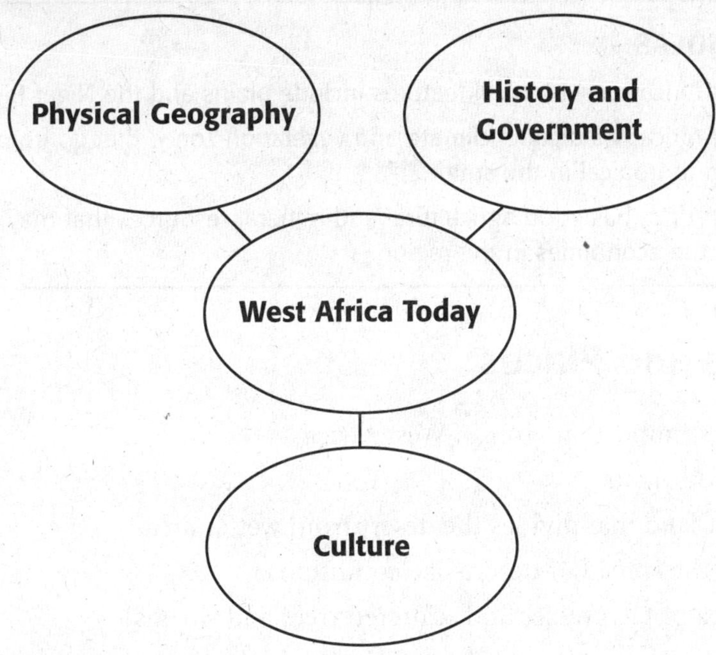

COMPREHENSION AND CRITICAL THINKING

Use the graphic organizer to answer the following questions.

1. Identify In which circle would you list details about ethnic groups?

____

2. Analyze How has the physical geography of the Sahel countries affected their economies?

____

____

3. Describe If "European Influences" was added to the "History" circle, what two major details could you add?

____

____

4. Make a Generalization How would you describe the effect of West African governments on the economies there?

____

____

West Africa

Section 1

MAIN IDEAS

1. West Africa's key physical features include plains and the Niger River.
2. West Africa has distinct climate and vegetation zones that go from arid in the north to tropical in the south.
3. West Africa has good agricultural and mineral resources that may one day help the economies in the region.

Key Terms and Places

Niger River most important river in West Africa

zonal organized by zone

Sahel a strip of land that divides the desert from wetter areas

desertification the spread of desert-like conditions

savanna an area of tall grasses and scattered trees and shrubs

Section Summary

PHYSICAL FEATURES

The main physical features in West Africa are plains and rivers. Most of the region is covered by plains. Plains along the coast have most of the region's cities. People on inland plains usually farm or raise animals. There are a few highlands in the southwest and northeast of the region.

The **Niger River** is the most important river in the region. It brings water to the people of the region for farming and fishing. It also provides a transportation route. It has an inland delta hundreds of miles from the coast where it divides into a network of channels, swamps, and lakes.

> Underline the sentences that describe the importance of the Niger River to the region.

CLIMATE AND VEGETATION

West Africa has four climate regions, which are **zonal**, or organized by zone. They stretch from east to west. The zone farthest north is part of the largest desert in the world, the Sahara.

Just to the south of the Sahara is a region called the **Sahel**. It is a strip of land that divides the

> Circle the name of the largest desert in the world.

Section 1, _continued_

desert from wetter areas. It has a steppe climate where rainfall varies greatly from year to year. Although it is very dry, enough plants grow there to support some grazing animals.

Because animals have overgrazed the Sahel and people have cut trees for firewood, the wind blows soil away. There has also been drought in the area. This has caused **desertification**, or the spread of desert-like conditions.

To the south of the Sahel is **savanna**, an area of tall grasses and scattered trees and shrubs. When rain falls regularly, it is a good area for farming.

The coasts of the Atlantic Ocean and the Gulf of Guinea have a humid tropical climate. Much rain there supports tropical forests. Many trees have been cut to make room for the growing population.

> **What are two causes of desertification?**
> _______________________
> _______________________

> **Why have many trees been cut from tropical rain forests?**
> _______________________
> _______________________

RESOURCES

Because of the good farmland and climate in some areas, agricultural products are an important resource. These include coffee, coconuts, peanuts, and cacao, which is used to make chocolate. West Africa also has minerals such as diamonds, gold, iron ore, and bauxite, which is the source of aluminum. Oil is the region's most valuable resource. Nigeria is a major exporter of oil, which is found near its coast.

> **List four mineral resources found in West Africa.**
> _______________________
> _______________________

CHALLENGE ACTIVITY

Critical Thinking: Identifying Cause and Effect Why do you think fewer people live in the northern portion of the region than in the southern portion? Write a paragraph to explain your reasoning.

       Interactive Reader and Study Guide

West Africa

Section 2

MAIN IDEAS

1. In West Africa's history, trade made great kingdoms rich, but this greatness declined as Europeans began to control trade routes.

2. The culture of West Africa includes many different ethnic groups, languages, religions, and housing styles.

Key Terms and Places

Timbuktu the cultural center of the Songhai Empire in the 1500s

animism the belief that bodies of water, animals, trees, and other natural objects have spirits

extended family a group of family members that includes the father, mother, children, and close relatives in one household

Section Summary

HISTORY

One of the earliest kingdoms in West Africa was Ghana, which became rich and powerful by about 800. It controlled trade in gold and salt across the Sahara. In about 1300, the empire of Mali took over. It controlled the trade routes and supported artists and scholars. The empire of Songhai took control in the 1500s. Its cultural center was **Timbuktu**. It had a university, mosques, and many schools. But invasions weakened it, and trade decreased.

Europeans began the Atlantic slave trade in the 1500s. Their colonies in the Americas needed labor to work on large plantations. European traders sold enslaved Africans to colonists. Families were split up, and many people died. By the time the slave trade ended in the 1800s, millions of Africans had been taken from their homes.

European countries claimed colonies in West Africa in the late 1800s and kept control until after World War II. They built schools, roads, and railroads, but many Africans gave up farming and worked for low wages. All the countries in West Africa became independent by 1974.

> Underline the sentence describing Timbuktu.

> When did the Atlantic slave trade begin? When did it end?
>
> _______________________
>
> _______________________

   Interactive Reader and Study Guide

CULTURE

The societies in West Africa have been influenced
by African cultures, European culture, and Islam.
There are hundreds of ethnic groups in the region.
European colonizers drew borders for countries
that put different ethnic groups in one country or
separated one group into different countries. Many
West Africans are more loyal to their own ethnic
groups than to their country. The groups speak
hundreds of languages. The use of colonial
languages or West African languages that many
people share helps with communication.

West African religions include Islam in the north
and Christianity in the south. Both were introduced
by traders coming to the area. Traditional religions
are forms of **animism**. Animism is the belief that
bodies of water, animals, trees, and other natural
objects have spirits.

Some people in the region wear Western-style
clothing. Others wear traditional cotton clothing
which is loose and flowing. Rural homes are small
and often circular. They are made from mud
or straw because there are few trees for wood.
Extended families often live close together in a
village. An **extended family** includes parents,
children, and other relatives. West Africa's cities
have modern buildings. Extended families may live
together in houses or high-rise apartments.

> Circle three influences on West African societies.

> What are three major religions practiced in West Africa?
> ________________________
> ________________________

> Why are many rural homes made from mud or straw?
> ________________________
> ________________________

CHALLENGE ACTIVITY

Critical Thinking: Comparing and Contrasting Think about the
diversity in West African societies and the diversity in U.S. society,
including religions, languages, and ethnic groups. Use a Venn diagram
to compare and contrast each society. Then write a paragraph that
summarizes the information in your chart.

West Africa

Section 3

MAIN IDEAS

1. Nigeria has many different ethnic groups, an oil-based economy, and one of the world's largest cities.
2. Most coastal countries of West Africa have struggling economies and weak or unstable governments.
3. Lack of resources in the Sahel countries is a main challenge to economic development.

Key Terms and Places

secede break away from the main country

Lagos the former capital of Nigeria and the most populous city in West Africa

famine an extreme shortage of food

Section Summary

NIGERIA

Nigeria is the second largest country in West Africa. It has the largest population in Africa and the region's strongest economy. The Igbo ethnic group unsuccessfully tried to **secede**, or break away from the main country, in the 1960s. Leaders moved the capital to Abuja where there are few people, partly to avoid ethnic conflicts. The government is now a democracy after the military ruled for many years.

Underline the name of the capital of Nigeria.

Nigeria's most important resource is oil. It accounts for 95 percent of the country's export earnings. The main industrial center is **Lagos**, which is the most populous city in West Africa and the former capital of Nigeria. Although Nigeria has many resources, many people are poor. It has a very high birthrate and cannot produce enough food. Corrupt government has also contributed to the country's poverty.

What are two causes of poverty in Nigeria?

____

____

OTHER COASTAL COUNTRIES

Small countries along the coast struggle to develop their economies and stabilize their governments. Senegal and Gambia produce peanuts and offer

    Interactive Reader and Study Guide

Section 3, *continued*

tourism sites. Guinea has some bauxite reserves. Guinea-Bissau has undeveloped mineral resources. Cape Verde is an island country with a democratic government.

Liberia was founded in the 1820s by Americans for freed slaves. A long civil war there ended in 2003. Civil war in Sierra Leone helped destroy the economies in the area and killed thousands. Ghana and Côte d'Ivoire have rich natural resources including gold, timber, and agricultural products, but these countries have also experienced civil war. Unstable governments and poor farming economies have hurt Togo and Benin.

> Circle the type of government Cape Verde has.

> What two coastal countries have the greatest natural resources?
> ________________________
> ________________________

SAHEL COUNTRIES

Drought and the expanding desert challenge the Sahel countries to feed their own people. Former nomads in Mauritania are now crowded into cities. Ethnic tensions continue to cause problems there. Niger has a very small amount of farmland where people grow staple, or main, crops. Drought and locusts created **famine**, or an extreme shortage of food, there in the early 2000s. Chad depends on farming and fishing in Lake Chad, although much of its water has evaporated in recent years. It began to export oil in 2004.

> Underline the causes of famine in Niger.

Much of Mali is desert with some farming in the south. It is one of the poorest countries in the world, but its economy is improving. Burkina Faso is also very poor and has few resources. Conflicts in the region have hurt its economy.

CHALLENGE ACTIVITY

Critical Thinking: Comparing and Contrasting Write an essay that compares and contrasts the government, resources, and economy of Nigeria with the other countries in the region.

East Africa

CHAPTER SUMMARY

	Tanzania and Kenya	**Rwanda and Burundi**	**Sudan and Uganda**	**The Horn**
Physical Geography	mountains, highlands, plains	highlands	highlands, plains	desert, valleys, highlands
History and Culture	many European settlers in Kenya, peaceful countries, English language	former German colonies, French language, Hutu and Tutsi ethnic groups	northern Sudan dominated by Arab Muslims, dictatorship in Uganda until recently, English spoken there	Ethiopia long independent, Eritrea recently independent, Somalis mostly Muslim, French and Arabic spoken in Djibouti
Economy and Resources	tourism, farming, diamonds and gold, geothermal energy	lack resources; rely on coffee and tea exports	farming, few mineral resources	farming, herding, tourism, trade
Challenges		ethnic conflicts, genocide	ethnic conflicts, genocide	drought, ethnic conflicts, civil war

COMPREHENSION AND CRITICAL THINKING

Use information from the graphic organizer to answer the following questions.

1. **Compare** What challenge is faced by most countries in the region?

____

____

2. **Draw Conclusions** Under which columns could you add the details *slave trade, imperialism, animist,* and *Swahili* for "History and Culture"?

____

____

3. **Summarize** Add a detail to the "Tanzania and Kenya" column for "Challenges" in the space provided.

East Africa

Section 1

> **MAIN IDEAS**
> 1. East Africa's physical features range from rift valleys to plains.
> 2. East Africa's climate is influenced by its location and elevation, and the region's vegetation includes savannas and forests.

Key Terms and Places

rift valley places on Earth's surface where the crust stretches until it breaks

Great Rift Valley the largest rift on Earth, made up of two rifts—the eastern rift and the western rift

Mount Kilimanjaro the highest mountain in Africa

Serengeti Plain one of Tanzania's largest plains, home to abundant wildlife

Lake Victoria Africa's largest lake and the source of the White Nile

drought period when little rain falls, and crops are damaged

Section Summary

PHYSICAL FEATURES

The landscape of East Africa is varied and a home to diverse and abundant wildlife. **Rift valleys** cut from north to south across the region. Rift walls are often steep cliffs that can rise as much as 6,000 feet. The **Great Rift Valley** is made up of two rifts.

East Africa has many volcanic mountains. The tallest of these is **Mount Kilimanjaro**. Although the mountain is located near the equator, its peak is covered with ice and snow. Another area of high elevation is the Ethiopian Highlands.

Some areas of East Africa are flat plains. The **Serengeti Plain** in Tanzania is one of the largest. Many kinds of wildlife live here, including elephants, giraffes, lions, and zebras. Tanzania established much of the plain as a national park.

A number of rivers and lakes are found in East Africa. The Nile is the world's longest river. It begins in East Africa. Then it flows north to the Mediterranean Sea. The source of the White Nile is **Lake Victoria**. The Blue Nile begins in the

> What is surprising about Mount Kilimanjaro?
> ________________________
> ________________________

> Underline examples of wildlife that can be found on the Serengeti Plain.

> Circle the name of the world's longest river.

Interactive Reader and Study Guide

Section 1, *continued*

Ethiopian Highlands. Both rivers meet in Sudan to form the Nile.

Lake Victoria is Africa's largest lake, but many lakes also lie along the rift valleys. Some of these lakes are extremely hot or salty.

> Underline the name of Africa's largest lake.

CLIMATE AND VEGETATION

East Africa has a variety of climate and vegetation. Latitude and elevation affect climate. For example, areas near the equator receive heavy rains. Farther from the equator, the weather is drier. When little rain falls, **droughts** can occur. During a drought, crops fail, cattle die, and people begin to starve. There have been severe droughts in East Africa.

> What happens during a drought?
> _______________________
> _______________________
> _______________________

The climate south of the equator is tropical savanna. In savannas, plants include tall grasses and scattered trees. The rift floors have grasslands and thorn shrubs.

Plateaus and mountains are found north of the equator. They have a highland climate and thick forests. The highlands receive a lot of rainfall. The mild climate makes farming possible. Many people live in the highlands. Forests are found at higher elevations.

> Circle the reasons that farming is possible in the highlands.

East of the highlands and on the Indian Ocean coast, the elevation is lower. Desert and steppe climates are found here. Vegetation is limited to shrubs and grasses.

CHALLENGE ACTIVITY

Critical Thinking: Making Generalizations Write a booklet for tourists to read before embarking on a helicopter tour of East Africa. What will they find most interesting about the region? Why?

Interactive Reader and Study Guide

East Africa

Section 2

MAIN IDEAS

1. The history of East Africa is one of religion, trade, and European influence.
2. East Africans speak many different languages and practice several different religions.

Key Terms and Places

Nubia part of Egypt and Sudan, an early center of Christianity in Africa

Zanzibar an East African island that became an international slave trading center in the late 1700s

imperialism a practice that tries to dominate other countries' government, trade, and culture

Section Summary

HISTORY

Early civilizations in East Africa were highly developed. Later, Christianity and Islam were brought to the region.

Ethiopia was an early center of Christianity. From there, Christianity spread to **Nubia**, in present-day Egypt and Sudan. An early Christian emperor in Ethiopia was Lalibela. He built 11 rock churches in the early 1200s.

> Circle the names of the two early centers for Christianity.

Muslim Arabs from Egypt brought Islam into northern Sudan. Islam also spread to the Indian Ocean coast of present-day Somalia.

> Who brought Islam to East Africa?
> ____________________
> ____________________

The East African slave trade dates back more than 1,000 years. In the 1500s the Portuguese built forts and towns on the coast. The island of **Zanzibar** became a center of the international slave trade. Later, Europeans built plantations with slave labor in the interior.

When most European nations ended slavery in the early 1800s, they shifted their focus to trading goods such as gold, ivory, and rubber. Soon after, the European powers divided up most of Africa. They used **imperialism** to keep power. This is a

> How did most Europeans control their colonies?
> ____________________
> ____________________

 Interactive Reader and Study Guide

Section 2, *continued*

policy of taking over other countries' governments, trade, and culture. The British controlled much of East Africa.

Large numbers of Europeans settled in Kenya. But most colonial rulers used African deputies to control the countries. Many deputies were traditional chiefs. They often favored their own peoples. This caused conflict between ethnic groups. These conflicts have made it hard for governments to influence feelings of national identity.

Most East African countries gained independence in the early 1960s. Ethiopia, however, was never colonized. Independence did not solve all the problems of the former colonies. New challenges faced the newly independent countries.

Underline the name of the only East African country that was never colonized.

CULTURE

East Africa's culture is diverse. For example, people speak many different languages. French is the official language in Rwanda, Burundi, and Djibouti. English is spoken in Uganda, Kenya, and Tanzania. Swahili is an African language spoken by about 80 million East Africans. Other languages include Amharic, Somali, and Arabic.

Circle the three main languages spoken in East Africa.

Religion and family traditions are important in East Africa. Religions vary within and among ethnic groups. Most of the cultures honor ancestors.

Many East Africans are followers of animist religions. They believe the natural world contains spirits. Some Africans also combine ancient forms of worship with Christianity and Islam.

CHALLENGE ACTIVITY

Critical Thinking: Analyzing Information Ask students to imagine that they are the leader of an East African country. Have them write a speech to be delivered at the country's one-year anniversary celebration.

East Africa

Section 3

MAIN IDEAS

1. National parks are a major source of income for Tanzania and Kenya.
2. Rwanda and Burundi are densely populated rural countries with a history of ethnic conflict.
3. Both Sudan and Uganda have economies based on agriculture, but Sudan has suffered from years of war.
4. The countries of the Horn of Africa are among the poorest in the world.

Key Terms and Places

safari an overland journey to view African wildlife

geothermal energy energy produced from the heat of Earth's interior

genocide the intentional destruction of a people

Darfur a region of Sudan

Mogadishu port city in Somalia

Section Summary

TANZANIA AND KENYA TODAY

Tanzania and Kenya rely on agriculture and tourism. Tourists come to see wildlife on **safaris**. Much of Kenya has been set aside as national parkland to protect wildlife.

Tanzania is rich in gold and diamonds. But most of its people are farmers. The mountains of Kenya provide rich soils for coffee and tea. **Geothermal energy**, another important resource, rises up through cracks in the rift valleys.

Dar es Salaam was once Tanzania's capital. Today the capital is Dodoma. Kenya's capital is Nairobi. In 1998, al Qaeda terrorists bombed the U.S. embassies in Dar es Salaam and Nairobi.

> Circle the capital cities of Tanzania and Kenya.

RWANDA AND BURUNDI TODAY

Rwanda and Burundi were once German colonies. Both have experienced conflicts between ethnic groups—the Tutsi and Hutu. This conflict led to **genocide** in Rwanda. Genocide is the intentional

> Underline the names of two ethnic groups in Rwanda and Burundi.

    Interactive Reader and Study Guide

Section 3, *continued*

destruction of a people. Rwanda and Burundi are densely populated. They export tea and coffee.

SUDAN AND UGANDA TODAY

Sudan is Africa's largest country. During the last several decades Muslims and Christians have fought a civil war. An Arab militia group has killed tens of thousands in a region of Sudan called **Darfur**.

Uganda was a military dictatorship for several decades. The country has become more democratic since 1986. Economic progress has been slow, however. About 80 percent of Ugandans work in agriculture. Coffee is the country's main export.

> **Circle the names of the groups fighting a civil war in Sudan.**

THE HORN OF AFRICA

The Horn of Africa is made up of four countries—Ethiopia, Eritrea, Somalia, and Djibouti.

Ethiopia has always been independent. Its people are mostly Christian and Muslim. Agriculture is the main economic activity. Severe droughts during the last 30 years have led to starvation. Plenty of rain has fallen in recent years, however.

Eritrea was once an Italian colony. Then it became part of Ethiopia. It broke away in 1993. Today tourism and farming are important to the growing economy.

Most Somalis are nomadic herders. Clans have fought over grazing rights. They have fought for control of port cities such as **Mogadishu**. Civil war and drought led to starvation in the 1990s.

Djibouti is a small desert country. It lies on a strait between the Red Sea and the Indian Ocean. Its port is a major source of income. Djibouti was once a French colony. Its people include the Issa and the Afar. They fought a civil war that ended in 2001.

> **What four countries make up the Horn of Africa?**
> _______________________
> _______________________
> _______________________

> **Underline important industries in Eritrea.**

CHALLENGE ACTIVITY

Critical Thinking: Drawing Conclusions Suppose you joined a United Nations delegation in East Africa. Propose ways the UN could help people in the region.

Central Africa

CHAPTER SUMMARY

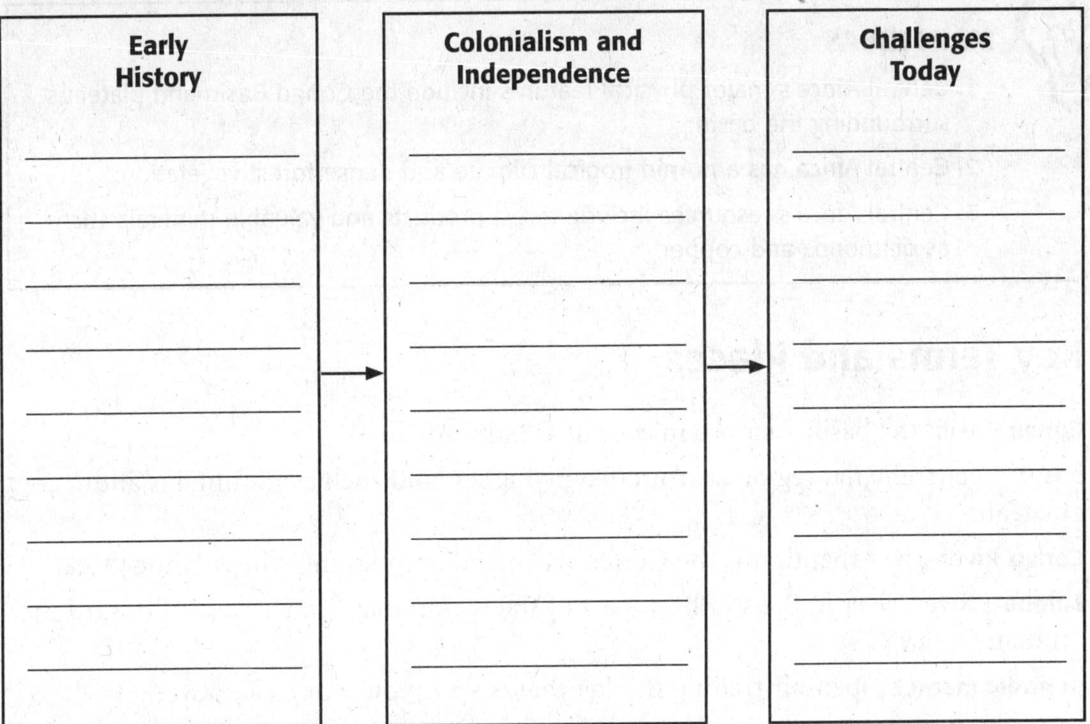

COMPREHENSION AND CRITICAL THINKING

Complete the graphic organizer by filling in information about the
countries of Central Africa. Use details from the graphic organizer to
examine how events in one period affected events or contributed to
problems in the period that followed it.

1. Describe In what ways are the ethnic groups of Central Africa different from
each other?

____

____

2. Explain How has the physical geography of Central Africa helped the region's
economy? How has it been an obstacle to economic growth?

____

____

3. Analyzing Information How have unstable or corrupt governments contributed
to the challenges facing the region today?

____

____

Central Africa

Section 1

MAIN IDEAS

1. Central Africa's major physical features include the Congo Basin and plateaus surrounding the basin.

2. Central Africa has a humid tropical climate and dense forest vegetation.

3. Central Africa's resources include forest products and valuable minerals such as diamonds and copper.

Key Terms and Places

Congo Basin the basin near the middle of Central Africa

basin a generally flat region surrounded by higher land such as mountains and plateaus

Congo River river that drains the Congo Basin and empties into the Atlantic Ocean

Zambezi River river in the southern part of the region that flows eastward toward the Indian Ocean

periodic market open-air trading market that is set up once or twice a week

copper belt area where copper is found that runs through northern Zambia and southern Democratic Republic of the Congo

Section Summary

PHYSICAL FEATURES

Central Africa lies between the Atlantic Ocean and the Western Rift Valley. Near the middle of the region is the **Congo Basin**. Plateaus and low hills surround the **basin**. The highest mountains in Central Africa are east of the basin along the Western Rift Valley. Lake Nyasa, also called Lake Malawi, and Lake Tanganyika lie along the rift. The **Congo River** is an important transportation route. It drains the Congo Basin and has hundreds of smaller rivers flowing into it. The many rapids and waterfalls prevent ships from traveling from the interior of the region all the way to the Atlantic. The **Zambezi River** flows toward the Indian Ocean. Many rivers in Angola and Zambia and from Lake Nyasa flow into the Zambezi. It also has many waterfalls, including Victoria Falls.

> **Where are the tallest mountains in the region found?**
> ______________________________
> ______________________________

CLIMATE, VEGETATION, AND ANIMALS

Because of its position along the equator, the Congo Basin and much of the Atlantic coast have a humid tropical climate with warm temperatures and plenty of rainfall all year. The warm, wet climate has led to the growth of dense tropical forests. These forests are home to such animals as gorillas, elephants, and okapis—relatives of the giraffe. However, large areas of these forests are now being cleared for farming and logging. To protect the forests and the animals that live there, some Central African governments have set up national park areas.

The climate north and south of the Congo Basin is a tropical savanna climate with warm weather all year but with distinct dry and wet seasons. Grasslands, scattered trees, and shrubs are the main vegetation. In the east, the high mountains have a highland climate. The far southern part of the region has dry steppe and desert climates.

> How has the warm and rainy climate of the Congo Basin affected the vegetation?
>
> _______________________
>
> _______________________

RESOURCES

The tropical climate is good for farming. Most people are subsistence farmers. Many grow crops for sale. In rural areas people sell goods at a **periodic market**. Other natural resources include timber from forests and rivers, which are important to travel, trade, and production of hydroelectricity. Some countries have oil, natural gas, and coal and valuable minerals such as copper, uranium, tin, zinc, diamonds, gold, and cobalt. Most of Africa's copper is in the **copper belt**. However, political problems and poor transportation have kept these resources from being fully developed.

> Name three kinds of natural resources in Central Africa.
>
> _______________________
>
> _______________________

> Circle the area where most of the copper is mined.

CHALLENGE ACTIVITY

Critical Thinking: Making Inferences Why might the Congo River be important to the development of the region's mineral resources? Write a brief paragraph to explain your answer.

Central Africa

Section 2

MAIN IDEAS

1. Great African kingdoms and European colonizers have influenced the history of Central Africa.

2. The culture of Central Africa includes many ethnic groups and languages, but it has also been influenced by European colonization.

Key Terms and Places

Kongo Kingdom one of the most important kingdoms in Central Africa, founded in the 1300s near the mouth of the Congo River

dialects regional varieties of a language

Section Summary

HISTORY

Early humans have lived in Central Africa for many thousands of years. About 2,000 years ago people from outside the region began to migrate into the region. They formed kingdoms. The **Kongo Kingdom** was one of the most important. Its people grew rich from trade in animal skins, shells, slaves, and ivory. They set up trade routes to western and eastern Africa.

> **How did the Kongo Kingdom become rich?**
> _______________________
> _______________________

The arrival of Europeans in the 1400s changed the region. The Europeans first came looking for trade goods. They wanted the region's forest products and other natural resources. Europeans also traded with some Central African kingdoms for slaves. The slave trade lasted about 300 years, with tragic effects for millions of enslaved Africans. They were forced to go to colonies in the Americas. At first some African kingdoms became richer by trading with Europeans, but in time these kingdoms were weakened by the Europeans.

> **Underline the sentence that tells why the Europeans came to the region.**

> **How did the slave trade affect the people of the region?**
> _______________________
> _______________________

In the late 1800s France, the United Kingdom, Belgium, Germany, Spain, and Portugal divided all of Central Africa into colonies. The colonial borders ignored the homelands of different groups and put groups with different languages and customs

　Interactive Reader and Study Guide

Section 2, *continued*

together. This led to serious problems and conflicts after these colonies became independent nations.

Central African colonies gained independence after World War II. In some cases they fought bloody wars to do so. Angola was the last colony to win independence. It did not become an independent country until 1975. After independence, fighting continued among ethnic groups within new countries. The Cold War led to more fighting. Both the Soviet Union and the United States supported different groups in small wars that killed many people.

> **What problems did colonial borders create for the people of the region?**
>
> _______________________
>
> _______________________

> **How did the United States and the Soviet Union add to the conflicts in Africa?**
>
> _______________________
>
> _______________________

CULTURE

People in Central Africa speak different languages. Many speak different **dialects** of Bantu languages. Most countries also have official languages such as French, English, Portuguese, or Spanish because of the influence of the European colonial powers. The region's colonial history also influenced religion. Many people in former French, Spanish, and Portuguese colonies are Roman Catholics. In former British colonies, many people are Protestants. The northern part of the region near the Sahel has many Muslims. Many Muslims and Hindus live in Zambia.

The traditional cultures of Central Africa's ethnic groups have influenced the arts. The region is famous for sculpture, carved wooden masks, and colorful cotton gowns. The region is also the birthplace of a popular musical instrument called the *likembe*, or thumb piano, and a type of dance music called *makossa*.

> **List four types of Central African arts.**
>
> _______________________
>
> _______________________
>
> _______________________
>
> _______________________

CHALLENGE ACTIVITY

Critical Thinking: Evaluating Information Do you think colonial rule helped or hurt the people of Central Africa? Explain your answer in a brief essay.

     Interactive Reader and Study Guide

Central Africa

Section 3

> **MAIN IDEAS**
>
> 1. The countries of Central Africa are mostly poor, and many are trying to recover from years of civil war.
> 2. Challenges to peace, health, and the environment slow economic development in Central Africa.

Key Terms and Places

Kinshasa capital of the Democratic Republic of the Congo

inflation the rise in prices that occurs when currency loses its buying power

malaria a disease spread by mosquitoes that causes fever and pain

malnutrition the condition of not getting enough nutrients from food

Section Summary

COUNTRIES OF CENTRAL AFRICA

Most Central African countries are very poor. Because of colonial rule and civil wars, they have had problems building stable governments and strong economies. Before independence, the Democratic Republic of the Congo was a Belgian colony. After the Belgians left, the country had few teachers or doctors. Ethnic groups fought for power. For several decades a corrupt dictator, Mobutu Sese Seko, ran the country. The economy collapsed, but Mobutu became very rich. In 1997 a civil war ended his rule. Although the country has resources such as minerals and forest products, civil war, bad government, and crime have scared away foreign businesses. Most people live in rural areas, but many are moving to the capital, **Kinshasa**.

Since independence, other Central African countries have had similar problems. Central African Republic has had military coups, corrupt leaders, and improper elections. Civil wars in the Republic of the Congo and Angola have hurt their governments and economies. In Angola land mines left from its civil war, high **inflation**, and corrupt officials have caused problems.

> Underline two reasons why Central African countries have had trouble building stable governments and strong economies.

> Circle the problems that have stopped foreign businesses from investing in the Democratic Republic of the Congo.

Section 3, *continued*

In most countries the majority of people are subsistence farmers. The economies of many countries depend heavily on the sale of natural resources like oil, copper, or diamonds or on export crops like coffee or cocoa. The Republic of the Congo has oil and forest products. Angola has diamonds and large oil deposits. Zambia's economy depends on copper mining. Malawi relies on farming and foreign aid. Oil discoveries in Equatorial Guinea and São Tomé and Príncipe may help their economies improve. Most nations need better railroads or ports for shipping goods to develop their natural resources.

Cameroon's stable government has helped its economy grow. It has good roads and railways. Gabon also has a stable government. Its economy is the strongest in the region. Half of its income comes from oil.

> What kind of work do most people in Central African countries do?

> What would help the countries of the region get more out of their natural resources?

ISSUES AND CHALLENGES

The region faces serious challenges from wars, diseases such as **malaria** and AIDS, and threats to the environment. Wars are one cause of the poor economies in the region. Deaths from wars and disease have resulted in fewer older, more skilled workers. Health officials and some national governments are trying to control malaria by teaching people ways to protect themselves. Other issues include rapid population growth, food shortages, and **malnutrition**. Because so many people die of disease, the region has a very young population. Farmers can't meet the demand for food, so food shortages occur. Other threats are the destruction of tropical forests and open-pit mining of diamonds and copper, which destroys the land.

> List one challenge facing the region today. Write a sentence describing one effect of this problem on the people or land of the region.

CHALLENGE ACTIVITY

Critical Thinking: Cause and Effect Create a cause-and-effect chart that lists possible effects of rapid population growth on unstable governments and poor economies. Then write a short paragraph that summarizes the information in your chart.

Interactive Reader and Study Guide

Southern Africa

CHAPTER SUMMARY

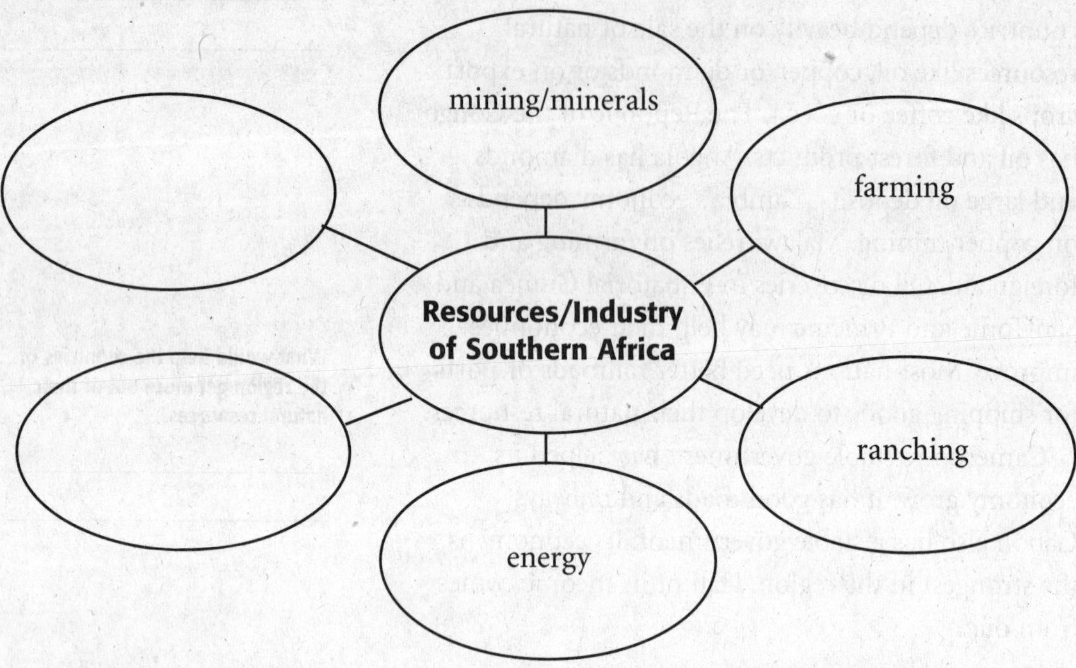

COMPREHENSION AND CRITICAL THINKING

Use information from the graphic organizer to answer the following questions.

1. Identify Which two of the industries above rely on fertile land?

____

____

2. Interpret Of the following five words (*manufacturing, sports, fishing, apartheid, ivory*), which two belong in the graphic organizer above? Write the two correct words in the empty circles.

____

____

3. Draw Conclusions Based on the chart above, what do you think is the most valuable resource in Southern Africa? Give reasons for your answer.

____

____

Southern Africa

Section 1

MAIN IDEAS

1. Southern Africa's main physical feature is a large plateau with plains, rivers, and mountains.
2. The climate and vegetation of Southern Africa is mostly savanna and desert.
3. Southern Africa has valuable mineral resources.

Key Terms and Places

escarpment the steep face at the edge of a plateau or other raised area

veld open grassland areas of South Africa

Namib Desert a desert located on the Atlantic coast, the driest place in the region

pans low, flat areas into which ancient streams drained and later evaporated

Section Summary

PHYSICAL FEATURES

Southern Africa is covered with grassy plains, steamy swamps, mighty rivers, rocky waterfalls, and steep mountains and plateaus.

Most of Southern Africa lies on a large plateau. The steep face at the edge of a plateau or other raised area is called an **escarpment**. In eastern South Africa, part of the escarpment is made up of a mountain range called the Drakensberg. Farther north, the Inyanga Mountains separate Zimbabwe and Mozambique.

> **What is an escarpment?**
> _______________________
> _______________________

Many large rivers cross Southern Africa's plains. The Okavango flows from Angola into a huge basin in Botswana. The Orange River passes through the Augrabies Falls and flows into the Atlantic Ocean.

> **Name two major rivers of Southern Africa.**
> _______________________
> _______________________

CLIMATE AND VEGETATION

Southern Africa's climates change from east to west. The east coast of the island of Madagascar is the wettest place in the region. In contrast to the eastern part of Africa, the west is very dry. Deserts along the Atlantic coast give way to plains with semiarid and steppe climates. Much of Southern

 Interactive Reader and Study Guide

Section 1, *continued*

Africa is covered by a large savanna region. On this grassland plain, shrubs and short trees grow. These grassland areas are known as the **veld** in South Africa.

The **Namib Desert** on the Atlantic Coast is the driest area in the region. The Kalahari Desert covers most of Botswana. Here ancient streams have drained into low, flat areas, or **pans**. On these pans, a glittering white layer forms when the streams dry up and leave minerals behind.

While the mainland is mostly dry, Madagascar has lush vegetation and tropical forests. Many animals, such as lemurs, are found here and nowhere else in the world. Unfortunately, rain forest destruction has endangered many of Madagascar's animals.

> Which desert covers most of Botswana?
>
> _______________________

RESOURCES

Rich in natural resources, Southern Africa has useful rivers, forests, and minerals. Its rivers provide a source of hydroelectric power and irrigation for farming. Forests are a source of timber. Mineral resources include gold, diamonds, platinum, copper, uranium, coal, and iron ore. Mining is very important to Southern Africa's economy. However, mining can harm the surrounding natural environments.

> What are some of Southern Africa's mineral resources?
>
> _______________________
>
> _______________________

CHALLENGE ACTIVITY

Critical Thinking: Summarizing Based on what you've read so far, write a one-sentence summary to go with each of the following headings.

a. Physical Features of Southern Africa

b. Climates of Southern Africa

c. Vegetation of Southern Africa

d. Resources of Southern Africa

Southern Africa

Section 2

MAIN IDEAS

1. Southern Africa's history began with hunter-gatherers, followed by great empires and European settlements.
2. The cultures of Southern Africa are rich in different languages, religions, customs, and art.

Key Terms and Places

Great Zimbabwe the stone-walled capital built by the Shona in the late 1000s

Cape of Good Hope area at the tip of Africa near where a trade station was set up by the Dutch in 1652

Afrikaners Dutch, French, and German settlers and their descendants living in South Africa

Boers Afrikaner frontier farmers who had spread out from the original Cape colony

apartheid the policy of racial separation set up by South Africa's government

township the separate areas where blacks had to live under apartheid

Section Summary

HISTORY

Bantu farmers moved from West Africa to Southern Africa about 2,000 years ago. In the 1700s, Europeans settled on the Southern Africa coast. They changed the region's landscape and way of life forever.

> How do you think Europeans changed the landscape and ways of life in the region?
>
> _______________________
>
> _______________________

The Khoisan peoples lived in Southern Africa for centuries. They were hunter-gatherers and herders. When the Bantu came to the area, they brought new languages and iron tools.

A Bantu group, the Shona, built an empire that reached its peak in the 1400s. The Shona farmed, raised cattle, and traded gold. They also built **Great Zimbabwe**, a stone-walled capital made of huge granite boulders and stone blocks. The city became a large trading center until the gold trade slowed.

> What group built an empire in Southern Africa?
>
> _______________________
>
> _______________________

In the late 1400s Portuguese traders set up bases on the Southern African coast. These bases served as a stopping place between Portugal and Asia. Other Europeans, including the Dutch, arrived after

    Interactive Reader and Study Guide

Section 2, *continued*

the Portuguese. In 1652 the Dutch set up a trade station in a natural harbor near the **Cape of Good Hope**. The Cape sits at the tip of Africa.

Other Europeans settled on the Cape. In South Africa, the Dutch, French, and German settlers and their descendants were called **Afrikaners**.

The British took over the Cape area in the early 1800s. The **Boers**, Afrikaner frontier farmers, tried to stop the British, but lost. At about this time, the Zulu became a powerful force in the region. The Zulu were a Bantu-speaking group. However, the British defeated them and took over this land, too.

When slavery ended in the British Empire in the 1830s, trade shifted from slaves to ivory. Then trade shifted away from ivory, to diamonds and gold, which were found in South Africa in the 1860s.

South Africa was ruled by white Afrikaners and became more racist in the 1900s. Black South Africans who opposed them formed the African National Congress (ANC). The white government set up a policy called **apartheid**, which divided people into four groups: whites, blacks, Coloureds, and Asians. Coloureds and Asians could only live in certain areas. Blacks had to live in separate areas, called **townships**. They had few rights.

Starting in the 1960s, the Southern Africa colonies slowly gained independence from Europe.

> **Who were the Boers?**
> _______________________
> _______________________

> **Why do you think the ivory trade ended?**
> _______________________
> _______________________

> **What were townships?**
> _______________________
> _______________________

CULTURE

Southern Africa has a rich and diverse culture. Its people belong to hundreds of different ethnic groups. They speak many languages, most of which are related to Khoisan or Bantu. They practice different religions, including Christianity and traditional African religions. Its arts reflect its many cultures, using traditional ethnic designs and crafts.

CHALLENGE ACTIVITY

Critical Thinking: Drawing Inferences What do you think the culture of Southern Africa would be like today if Europe still ruled over the region?

Southern Africa

Section 3

MAIN IDEAS

1. South Africa ended apartheid and now has a stable government and strong economy.
2. Some countries of Southern Africa have good resources and economies, but several are still struggling.
3. Southern African governments are responding to issues and challenges such as drought, disease, and environmental destruction.

Key Terms and Places

sanctions economic or political penalties imposed by one country on another to force a change in policy

Cape Town city in South Africa that attracts many tourists

enclave a small territory surrounded by a foreign territory

Section Summary

SOUTH AFRICA

Today South Africa has made great progress, but challenges remain. Perhaps South Africa's biggest challenge has been ending apartheid. Many people objected to apartheid. As a result, some countries put **sanctions** —penalties to force a change in policy—on South Africa. Protest within the country increased as well. In response, the government outlawed the African National Congress (ANC), a group defending the rights of black South Africans.

> **What are sanctions?**
> _______________________
> _______________________

> **What is the ANC?**
> _______________________
> _______________________

In the late 1980s, South Africa moved away from apartheid. In 1990 the government released its political prisoners, including Nelson Mandela. He was elected South Africa's president in 1994. South Africa's new government is a republic. Its constitution stresses equality and human rights.

> **Who was elected president of South Africa in 1994?**
> _______________________
> _______________________

South Africa has the region's strongest economy, with more resources and industry than most African countries. Large cities such as Johannesburg and **Cape Town** contribute to the economy.

 Interactive Reader and Study Guide

OTHER COUNTRIES OF SOUTHERN AFRICA

Surrounded by South Africa, Lesotho and Swaziland are both enclaves. An **enclave** is a small territory surrounded by a foreign territory. Both countries are kingdoms, but are governed by an elected prime minister and a parliament.

Namibia gained independence in 1990. It is a republic. Most of its income comes from mineral resources. Fishing and ranching are also important.

Botswana is rich in mineral resources and has a stable, democratic government. Cattle ranching and diamond mining are its main economic activities.

Zimbabwe has a poor economy. It is also politically unstable. In 2000, the president began a land reform program, taking land from white farmers and giving it to black residents. However, food shortages resulted.

Mozambique is one of the world's poorest countries. The economy was hurt by a civil war. It relies on taxes collected on products shipped out of its ports from the interior of Africa.

Madagascar has an elected president, but the economy is struggling. The country is popular with tourists because of its unique plants and animals. Comoros is made up of four tiny islands. It is poor and politically unstable. However, the government hopes to improve education and promote tourism.

> **Compare the economies of Botswana and Zimbabwe.**
>
> _______________________
> _______________________
> _______________________
> _______________________

> **What does Mozambique's economy rely on?**
>
> _______________________
> _______________________

> **What makes Madagascar a tourist destination?**
>
> _______________________
> _______________________

ISSUES AND CHALLENGES

Southern Africa is doing better than most other African countries. Still the region faces many challenges, especially poverty, disease, and environmental destruction. The African Union (AU) is working to promote cooperation among African countries to try to solve these problems.

> **What is the AU?**
>
> _______________________
> _______________________

CHALLENGE ACTIVITY

Critical Thinking: Drawing Inferences Of the Southern African countries you've read about, how could some use their resources differently to help improve their economies? Give support for your answer.

History of Ancient India

CHAPTER SUMMARY

Aryan invasion of Indus and Ganges	led to	**development of the caste system**
unification of India into empires	led to	**stability and prosperity**
development of religion	led to	**changes in roles of early Indian kings**
stability of early Indian empires	led to	**advances in arts and sciences**

COMPREHENSION AND CRITICAL THINKING

Use information from the graphic organizer to answer the following questions.

1. **Recall** Which group introduced and developed the caste system in India?

   ____

   ____

2. **Identify Cause and Effect** Why does the unification of civilizations usually lead to prosperity and stability?

   ____

   ____

3. **Evaluate** Why do you think religion played such an important role in the way rulers were regarded in early civilizations?

   ____

   ____

4. **Draw a Conclusion** What characteristic of a civilization usually provides a sound basis for advances in arts and sciences?

   ____

   ____

History of Ancient India

Section 1

MAIN IDEAS

1. Located on the Indus River, the Harappan civilization also had contact with people far from India.
2. Harappan achievements included a writing system, city planning, and art.
3. The Aryan invasion changed India's civilization.

Key Terms and Places

Indus River major river in India along which the Harappan civilization developed

Harappa city in ancient India

Mohenjo Daro city in ancient India

Sanskrit the most important language of ancient India

Section Summary

HARAPPAN CIVILIZATION

India's first civilization was the Harappan civilization, which developed along the **Indus River**. Archaeologists believe Harappan civilization thrived between 2300 and 1700 BC. Harappan settlements were scattered over a huge area, but most lay next to rivers. The largest settlements were **Harappa** and **Mohenjo Daro**. The Harappans may have traded with people as far away as southern India and Mesopotamia.

HARAPPAN ACHIEVEMENTS

The Harappans developed India's first writing system. Although archaeologists have found examples of their writing, scholars have not been able to read it. Most information about Harappans comes from studying the ruins of cities, especially Harappa and Mohenjo Daro. These cities were well-planned and advanced. Each city was built in

> **Why do you think we know so little about the Harappans?**
>
> _______________________
>
> _______________________
>
> _______________________

Section 1, *continued*

the shadow of a fortress that could easily oversee the city streets. The streets themselves were built at right angles and had drainage systems. The Harappans also developed beautiful artisan crafts, some of which have helped historians draw conclusions about Harappan society. Harappan civilization collapsed by the early 1700s BC, possibly due to invasions or natural disasters.

> Why was it an advantage for the streets of Mohenjo Daro and Harappa to be viewed from a fortress?
>
> ___________________________
> ___________________________
> ___________________________
> ___________________________

ARYAN MIGRATION

Originally from Central Asia, the Aryans first reached India in the 2000s BC. Over time they spread south and east into central India and eventually into the Ganges River Valley. Much of what is known about the Aryans comes from a collection of religious writings called the Vedas.

Unlike the Harappans, Aryans lived in small communities run by a local leader, or raja. Aryan groups fought each other as often as they fought outsiders.

The Aryans spoke **Sanskrit** and memorized poems and hymns that survived by word of mouth. People later figured out how to write in Sanskrit. Sanskrit records are a major source of information about Aryan society. Today Sanskrit is the root of many modern South Asian languages.

> The early Aryans had a rich and expressive language, but they did not write. How did they preserve their poems and their history without writing?
>
> ___________________________
> ___________________________
> ___________________________
> ___________________________

CHALLENGE ACTIVITY

Critical Thinking: Drawing Inferences Write a short essay explaining what you think might have happened to the Harappan civilization.

Interactive Reader and Study Guide

History of Ancient India

Section 2

MAIN IDEAS

1. Indian society divided into distinct groups.
2. The Aryans formed a religion known as Brahmanism.
3. Hinduism developed out of Brahmanism and influences from other cultures.
4. The Jains reacted to Hinduism by breaking away.

Key Terms

caste system a division of Indian society into groups based on a person's birth, wealth, or occupation

reincarnation the belief that the soul, once a person dies, is reborn in a new body

karma the effects that good or bad actions have on a person's soul

nonviolence the avoidance of violent actions

Section Summary

INDIAN SOCIETY DIVIDES

Aryan society was divided into social classes. There were four main groups, called *varnas*. The Brahmins (BRAH-muhns) were priests and were the highest ranking varna. The Kshatriyas (KSHA-tree-uhs) were rulers or warriors. The Vaisyas (VYSH-yuhs) were commoners, including farmers, craftspeople, and traders. The Sudras (SOO-drahs) were laborers and servants.

> Rank the main groups of the Aryan social classes in order of importance, with one (1) being highest and four (4) being the lowest:
> Brahmins
> Sudras
> Kshatriyas
> Vaisyas

Eventually a more complex **caste system** developed, dividing Indian society into many groups based on birth, wealth, or occupation. Castes were family based. If you were born into a caste, you would probably stay in it for your whole life. Life for the lower castes was difficult, but those who had no caste, called untouchables, were ostracized.

> In ancient India, why was it important to belong to some caste?
>
> _______________________________
>
> _______________________________
>
> _______________________________

BRAHMANISM

Because Aryan priests were called Brahmins, the Aryan religion became known as Brahmanism. Brahmanism was perhaps the most important part

142

Interactive Reader and Study Guide

of ancient Indian life, as shown by the high status of the priest caste. The religion was based on the four Vedas, writings that contained ancient sacred hymns and poems. Over time, Aryan Brahmins and scholars wrote their thoughts about the Vedas. These thoughts were compiled into Vedic texts. The texts described rituals, such as how to perform sacrifices, and offered reflections from religious scholars.

HINDUISM DEVELOPS

Hinduism is India's largest religion today. It developed from Brahmanism and other influences. Hindus believe that there are many gods, but all gods are part of a universal spirit called Brahman. Hindus believe everyone has a soul, or atman, and the soul will eventually join Brahman. This happens when the soul recognizes that the world we live in is an illusion. Hindus believe this understanding takes several lifetimes, so **reincarnation**, or rebirth, is necessary. How you are reborn depends upon your **karma**, or the effects of good or bad actions on your soul. In the caste system, those who have good karma are born to higher castes. Those with bad karma are born into lower castes or maybe even an animal.

> **What is the Hindu name for the soul?**
>
> _______________________

> **Think about why you believe that the real world actually exists. Do you think you can prove that it does?**
>
> _______________________

JAINS REACT TO HINDUISM

The religion of Jainism developed in reaction to Hinduism. Jains believe in injuring no life, telling the truth, not stealing, and not owning property. Jains also practice **nonviolence**, or *ahimsa*. This emphasis on nonviolence comes from the belief that everything in nature is part of the cycle of rebirth.

CHALLENGE ACTIVITY

Critical Thinking: Drawing Inferences Do *ahimsa*, reincarnation, or karma have relevance in our society today? Pick one of these terms and write a one-page essay on how it may or may not be important in your life.

Section 3

MAIN IDEAS

1. Siddhartha Gautama searched for wisdom in many ways.

2. The teachings of Buddhism deal with finding peace.

3. Buddhism spread far from where it began in India.

Key Terms

fasting going without food

meditation the focusing of the mind on spiritual ideas

nirvana a state of perfect peace

missionaries people who work to spread their religious beliefs

Section Summary

SIDDHARTHA'S SEARCH FOR WISDOM

Not everyone in India accepted Hinduism. In the late 500s BC, a major new religion began to develop from questions posed by a young prince named Siddhartha Gautama (si-DAHR-tuh GAU-tuh-muh). Siddhartha was born to a wealthy family and led a life of comfort, but he wondered at the pain and suffering he saw all around him. By the age of 30, Siddharta left his home and family to look for answers about the meaning of life. He talked to many priests and wise men, but he was not satisfied with their answers.

Siddhartha did not give up. He wandered for years through the forests trying to free himself from daily concerns by **fasting** and **meditating**. After six years, Siddhartha sat down under a tree and meditated for seven weeks. He came up with an answer to what causes human suffering. Suffering is caused by wanting what one does not have, wanting to keep what one likes and already has, and not wanting what one dislikes but has. He began to travel and teach his ideas, and was soon called the Buddha, or "Enlightened One." From his teachings sprang the religion Buddhism.

Why did Siddhartha leave his life of luxury?

____

____

____

____

Can you think of a form of human suffering not covered by one of Siddhartha's three categories? If so, state briefly what it is.

____

____

____

____

TEACHINGS OF BUDDHISM

Buddhism is based upon the Four Noble Truths. These truths are: Suffering and unhappiness are part of life; suffering stems from our desire for pleasure and material goods; people can overcome their desires and reach **nirvana**, a state of perfect peace, which ends the cycle of reincarnation; and people can follow an eightfold path to nirvana, overcoming desire and ignorance.

These teachings were similar to some Hindu concepts, but went against some traditional Hindu ideas. Buddhism questioned the need for animal sacrifice. It also challenged the authority of the Brahmins. The Buddha said that each individual could reach salvation on his or her own. Buddhism also opposed the caste system.

> What is the name of the central teachings of Buddhism?
>
> _______________________
>
> _______________________

> Buddhist texts often refer to "the compassionate Buddha." Why is this term appropriate?
>
> _______________________
>
> _______________________
>
> _______________________

BUDDHISM SPREADS

Buddhism spread quickly throughout India. With the help of Indian king Asoka, Buddhist **missionaries** were sent to other countries to teach their religious beliefs. Missionaries introduced Buddhism to Sri Lanka and other parts of Southeast Asia, as well as Central Asia and Persia. It eventually spread to China, Japan, and Korea. In modern times, Buddhism has become a major global religion.

CHALLENGE ACTIVITY

Critical Thinking: Drawing Inferences Could you leave your family, home, and everything you know to preach what you believe to be a spiritual truth? You are preparing to follow the Buddha. Write a goodbye letter to your family explaining why you have chosen this life of sacrifice.

Section 4

MAIN IDEAS

1. The Mauryan Empire unified most of India.
2. Gupta rulers promoted Hinduism in their empire.

Key Terms

mercenaries hired soldiers

edicts laws

Section Summary

MAURYAN EMPIRE UNIFIES INDIA

Under Aryan rule, India was divided into several states with no central leader. Then, during the 300s BC, the conquests of Alexander the Great brought much of India into his empire. An Indian military leader named Candragupta Maurya followed Alexander's example and seized control of the entire northern part of India, using an army of **mercenaries**, or hired soldiers. The Mauryan Empire lasted for about 150 years.

Candragupta's complex government included a huge army and a network of spies. He taxed the population heavily for the protection he offered. Eventually, Candragupta became a Jainist monk and gave up his throne to his son. His family continued to expand the Indian empire.

Candragupta's grandson, Asoka, was the strongest ruler of the Mauryan dynasty. The empire thrived under his rule. But at last, tired of killing and war, Asoka converted to Buddhism. He sent Buddhist missionaries to other countries and devoted the rest of his rule to improving the lives of his people. He had workers build wells, tree-shaded roads, and rest houses, and raised large stone pillars carved with Buddhist **edicts**, or laws. When Asoka died, however, his sons struggled for power and foreign invaders threatened the country. The

Who inspired Indian leader Candragupta Maurya to unify India for the first time?

____

____

What is the relationship between Candragupta's government and the heavy taxes?

____

____

Asoka is sometimes regarded as proof that national security can coexist with peace. Do you think a leader like Asoka could be effective in the world today? Why or why not?

____

____

____

____

Mauryan Empire fell in 184 BC, and India remained divided for about 500 years. The spread of Buddhism steadily increased, while Hinduism declined.

GUPTA RULERS PROMOTE HINDUISM

A new dynasty was established in India. During the 300s AD, the Gupta Dynasty once again rose to unite and build the prosperity of India. Not only did the Guptas control India's military, they were devout Hindus and encouraged the revival of Hindu traditions and writings. The Guptas, however, also supported Jainism and Buddhism.

Indian civilization reached a high point under Candra Gupta II. He poured money and resources into strengthening the country's borders, as well as promoting the arts, literature, and religion.

The Guptas believed the caste system supported stability. This was not good for women, whose role under the empire was very restricted. Women were expected to marry, in weddings arranged by their parents, and raise children. A woman had to obey her husband and had few rights.

The Gupta Dynasty lasted until fierce attacks by the Huns from Central Asia during the 400s drained the empire of its resources. India broke up once again into a patchwork of small states.

CHALLENGE ACTIVITY

Critical Thinking: Drawing Inferences Asoka was strongly influenced by Buddhism. Candra Gupta II followed Hinduism. Write a short essay explaining which king you think was a better leader. How did their religion affect their rule? Keep in mind the situation of Indian society under both kings' reign.

History of Ancient India

Section 5

MAIN IDEAS

1. Indian artists created great works of religious art.

2. Sanskrit literature flourished during the Gupta period.

3. The Indians made scientific advances in metalworking, medicine, and other sciences.

Key Terms

metallurgy the science of working with metals

alloys mixtures of two or more metals

Hindu-Arabic numerals the numbering system invented by Indian mathematicians and brought to Europe by Arabs; the numbers we use today

inoculation a method of injecting a person with a small dose of a virus to help him or her build up defenses to a disease

astronomy the study of stars and planets

Section Summary

RELIGIOUS ART

Both the Mauryan and Gupta empires unified India and created a stable environment where artists, writers, scholars, and scientists could thrive. Their works are still admired today. Much of the Indian art from this period was religious, inspired by both Hindu and Buddhist teachings. Many beautiful temples were built during this time and decorated with elaborate wood and stone carvings.

> What was the main inspiration for art and literature during the Mauryan and Gupta empires?
>
> ___________________________
>
> ___________________________

SANSKRIT LITERATURE

Great works of literature were written in Sanskrit, the ancient Aryan language, during the Gupta Dynasty. The best-known works are the *Mahabharata* (muh-HAH-BAH-ruh-tuh) and the *Ramayana* (rah-MAH-yuh-nuh). The *Mahabharata*, a long story about the struggle between good and evil, is considered a classic Hindu text. The most famous passage is called the *Bhagavad Gita* (BUG-uh-vuhd GEE-tah). The *Ramayana* is the story of the Prince

> Sanskrit literature had a long tradition before it was written down. How were these early works first preserved?
>
> ___________________________
>
> ___________________________

Section 5, *continued*

Rama, a human incarnation of one of the three
major Hindu gods, Vishnu, who fights demons and
marries the beautiful princess Sita.

SCIENTIFIC ADVANCES

Scientific and scholarly work also blossomed during
the early Indian empires. Most prominent was the
development of **metallurgy**, the science of working
with metals. Indian technicians and engineers made
strong tools and weapons. They also invented
processes for creating **alloys**. Alloys, such as steel or
bronze, may be stronger or more useful than pure
metals like iron or copper.

> **What major modern industry involves the production of a widely used alloy?**
> _______________________
> _______________________

The numbers we use today, called **Hindu-Arabic numerals**, were first developed by Indian
mathematicians. They also created the concept of
zero, upon which all modern math is based.

> **What mathematical concept expresses the idea of "none?"**
> _______________________
> _______________________

Other sciences also benefited from this period
of Indian history. In medicine, Indians developed
the technique of **inoculation**, which is injecting a
person with a small dose of a virus to help him or
her build up defenses to a disease. Doctors could
even perform certain surgeries. India's fascination
with **astronomy**, the study of stars and planets, led
to the discovery of seven of the planets in our solar
system.

> **Indians at this period did not have telescopes. How do you think they discovered planets?**
> _______________________
> _______________________
> _______________________
> _______________________

CHALLENGE ACTIVITY

Critical Thinking: Drawing Inferences Our modern society borrows
significantly from the scientific and mathematical achievements of the
early Indian empires. Write a short play, story, or essay describing how
our modern world might look without these inventions.

History of Ancient China

CHAPTER SUMMARY

THE EARLY DYNASTIES

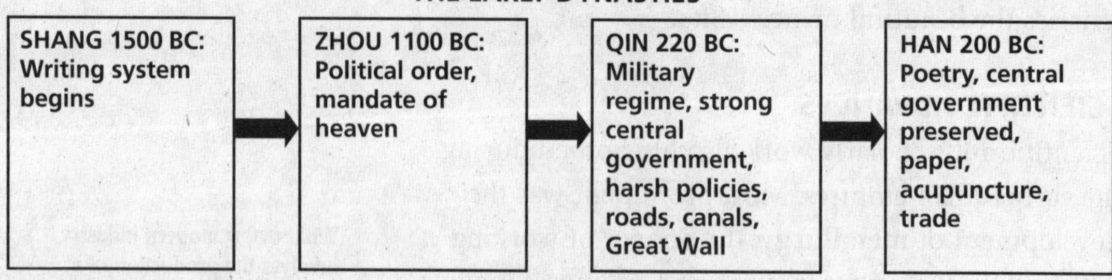

COMPREHENSION AND CRITICAL THINKING

Use information from the graphic organizer to answer the following
questions.

1. Identify Which of the early dynasties lasted the longest?

____

____

2. Draw Inferences Which dynasty improved on a rudimentary system of
communication that had probably existed for 2000 years?

____

____

3. Evaluate Which dynasty kept some beneficial aspects of the former regime and
ended its harsher aspects?

____

____

History of Ancient China

Section 1

MAIN IDEAS

1. Chinese civilization began along two rivers.
2. The Shang dynasty was the first known dynasty to rule China.
3. The Zhou and Qin dynasties changed Chinese society and made great advances.

Key Terms and Places

Chang Jiang a river in China

Huang He a river in China

mandate of heaven the idea that heaven chose China's ruler and gave him or her power

Xi'an present name of the capital city of the Qin dynasty

Great Wall a barrier that linked earlier walls that stood near China's northern border

Section Summary

CHINESE CIVILIZATION BEGINS

Like other ancient peoples, people in China first settled along rivers. By 7000 BC farmers grew rice in the **Chang Jiang** Valley. Along the **Huang He**, they grew millet and wheat. Some villages along the Huang He grew into large towns. Many artifacts were left in these towns, including pottery and tools. As Chinese culture became more advanced, people started to use potter's wheels and dig wells for water. Population continued to grow and villages spread into northern and southeastern China.

> Where did the Chinese first grow rice?
>
> ______________________
>
> ______________________

SHANG DYNASTY

Societies along the Huang He grew larger and more complex. The first dynasty for which we have clear evidence is the Shang. It was firmly established by the 1500s BC. The Shang made many advances, including China's first writing system. The Chinese symbols that are used today are based on those of the Shang period.

> Which dynasty provided the basis for China's writing system?
>
> ______________________
>
> ______________________

 Interactive Reader and Study Guide

Section 1, *continued*

Shang artisans made beautiful bronze containers for cooking and religious ceremonies. They also made ornaments, knives, and axes from jade. Shang astrologers developed a calendar based on the cycles of the moon.

ZHOU AND QIN DYNASTIES

The Zhou overthrew the Shang dynasty during the 1100s BC. The Zhou believed in the **mandate of heaven**, or the idea that they had been chosen by heaven to rule China. A new political order was established under the Zhou, with the emperor granting lands to lords in return for loyalty and military assistance. Peasants were below the lords, and owned little land. In 771 BC, the emperor was overthrown and China broke apart into many kingdoms, entering an era called the Warring States period.

The Warring States period ended when one state, the Qin, defeated the other states. In 221 BC the Qin king was able to unify China. He gave himself the title Shi Huangdi, which means "first emperor."

Shi Huangdi greatly expanded the size of China. He took land away from the lords and forced noble families to move to his capital, present-day **Xi'an**. Qin rule brought other changes to China. Shi Huangdi set up a uniform system of law. He also standardized the written language, and a new monetary system. The completion of the **Great Wall** was a major Qin achievement. The Qin built the wall to protect China from northern invaders.

Although Shi Huangdi unified China, no strong rulers took his place. China began to break apart once again within a few years of his death.

> Who was at the top of the Zhou political system? Who was at the bottom?
>
> _______________________
>
> _______________________

> List three ways that Shi Huangdi unified China.
>
> _______________________
>
> _______________________

CHALLENGE ACTIVITY

Critical Thinking: Drawing Inferences Using library or online resources, study examples of ancient Chinese writing. Use some of these symbols to illustrate something you have learned about China.

History of Ancient China

Section 2

> **MAIN IDEAS**
> 1. Han dynasty government was largely based on the ideas of Confucius.
> 2. Han China supported and strengthened family life.
> 3. The Han made many achievements in art, literature, and learning.

Key Terms

sundial a device that uses the position of shadows cast by the sun to tell the time of day

seismograph a device that measures the strength of earthquakes

acupuncture the practice of inserting fine needles through the skin at specific points to cure disease or relieve pain

Section Summary

HAN DYNASTY GOVERNMENT

Liu Bang (lee-oo bang), a peasant, led the army that won control of China after the collapse of the Qin dynasty. He earned the people's loyalty and trust. He lowered taxes for farmers and made punishments less severe. He set up a government that built on the foundation begun by the Qin. Liu Bang's successor, Wudi (woo-dee), made Confucianism the official government philosophy of China. To get a government job, a person had to pass a test based on Confucian teachings. However, wealthy and influential families still controlled the government.

> **Which feature of the Qin dynasty did the Han preserve?**
> _______________________
> _______________________

FAMILY LIFE

A firm social order took hold during Han rule. In the Confucian view, peasants made up the second-highest class. Merchants occupied the lowest class because they merely bought and sold what others had made. However, this social division did not indicate wealth or power. Peasants were still poor and merchants were still rich.

During Wudi's reign, Confucian teachings about the family were also honored. Children were taught from birth to respect their elders. Within the

> **Why does Confucian thinking devalue merchants?**
> _______________________
> _______________________
> _______________________

Section 2, *continued*

family, the father had absolute power. Han officials believed that if the family was strong and people obeyed the father, then people would obey the emperor, too. Chinese parents valued boys more highly than girls. Some women, however, still gained power. They could influence their sons' families. An older widow could even become the head of the family.

> Who had absolute power in the family under the Han?
> ___________________________
> ___________________________

> Circle the sentence that explains which women could become heads of families.

HAN ACHIEVEMENTS

The Han dynasty was a time of great accomplishments. Art and literature thrived, and inventors developed many useful devices. Han artists painted portraits and realistic scenes that showed everyday life. Poets developed new styles of verse. Historian Sima Qian wrote a complete history of China until the Han dynasty.

The Han Chinese invented paper. They made it by grinding plant fibers into a paste and then letting it dry in sheets. They made "books" by pasting sheets together into a long sheet that was rolled into a scroll.

> Name the greatest and most far-reaching Han invention.
> ___________________________
> ___________________________

Other Han innovations included the **sundial** and the **seismograph**. They developed the distinctive Chinese medical practice of **acupuncture** (AK-yoo-punk-cher). These and other Han inventions and advances are still used today.

CHALLENGE ACTIVITY

Critical Thinking: Drawing Inferences Could the Han dynasty have flourished if the Qin had not set up a strong government structure? Write a brief essay presenting your point of view on this question.

 Interactive Reader and Study Guide

History of Ancient China

Section 3

MAIN IDEAS

1. After the Han dynasty, China fell into disorder but was reunified by new dynasties.
2. Cities and trade grew during the Tang and Song dynasties.
3. The Tang and Song dynasties produced fine arts and inventions.

Key Terms and Places

Grand Canal a canal linking northern and southern China

Kaifeng capital of the Song dynasty

porcelain a thin, beautiful pottery invented by the Chinese

woodblock printing a form of printing in which an entire page is carved into a block of wood that is covered with ink and then pressed against paper to make a copy of the page

gunpowder a mixture of powders used in guns and explosives

compass an instrument that uses the earth's magnetic field to indicate direction

Section Summary

DISORDER AND REUNIFICATION

China broke apart into several kingdoms after the fall of the Han dynasty. This time period, sometimes known as the Period of Disunion, ended with the rise of the Sui dynasty in 589. Around this time, work was soon started on the **Grand Canal**, a system of waterways linking northern and southern China. The Sui dynasty, which did not last long, was followed by the Tang dynasty, which lasted nearly 300 years. This period was considered a golden age for China, with military reform, new law codes, and advances in art. The Song dynasty followed the Tang dynasty after a short period of disorder. The Song, like the Tang, ruled for about 300 years, and brought about many great achievements.

> **Under what dynasty was work on the Grand Canal begun?**
>
> _______________________
>
> _______________________

CITIES AND TRADE

Chinese cities grew and flourished as the trade centers of the Tang and Song dynasties. Chang'an (chahng-AHN), with a population of more than a

 Interactive Reader and Study Guide

million people, was by far the largest city in the
world at the time. Several other cities, including
Kaifeng, the Song capital, had about a million
people. Traders used the Grand Canal, a series of
waterways that linked major cities, to ship goods
and agricultural products throughout China.

Foreign trade used both land routes and sea
routes. China's Pacific ports were open to foreign
traders. Chinese exports included tea, rice, spices,
and jade. Especially prized by foreigners, however,
were silk and **porcelain**. The method of making
silk was kept secret for centuries.

> Why do you think the Chinese did not want foreigners to know how to make silk?
>
> _______________________
>
> _______________________
>
> _______________________

ARTS AND INVENTIONS

The Tang dynasty produced some of China's greatest
artists and writers, including the poets Li Bo and
Du Fu, and the Buddhist painter Wu Daozi
(DOW-tzee). The Song dynasty produced Li
Qingzhao (ching-ZHOW), perhaps China's greatest
female poet. Artists of the Tang and Song dynasties
created exquisite objects in clay, particularly porcelain items with a pale green glaze called celadon
(SEL-uh-duhn).

> Use the Internet or a library to find a poem by Li Bo.

The Tang and Song dynasties produced some of
the most remarkable—and important—inventions
in human history. The world's oldest-known
printed book, using **woodblock printing**, was
printed in China in 868. Later, during the Song
dynasty, the Chinese invented movable type for
printing. The Song dynasty also introduced the
world's first paper money. Two other inventions
include **gunpowder** and the **compass**.

> What printing technology ultimately superseded woodblock printing?
>
> _______________________
>
> _______________________

CHALLENGE ACTIVITY

Critical Thinking: Drawing Inferences Create a document showing an
exchange of goods between a Song dynasty Chinese trader and a foreign
merchant.

History of Ancient China

Section 4

> **MAIN IDEAS**
> 1. Confucianism, based on Confucius's teachings about proper behavior, dramatically influenced the Song system of government.
> 2. Scholar-officials ran China's government during the Song dynasty.

Key Terms

bureaucracy body of unelected government officials
civil service service as a government official
scholar-official an educated member of the government

Section Summary

CONFUCIANISM

Confucianism is the name given to the ideas of the Chinese philosopher Confucius. Confucius's teachings focused on ethics, or proper behavior, of individuals and governments. He argued that society would function best if everyone followed two principles, *ren* and *li*. *Ren* means concern for others, and *li* means appropriate behavior. Order in society is maintained when people know their place and behave appropriately.

For a thousand years after his death, Confucius's ideas went in and out of favor several times. Early in the Song dynasty, however, a new version of Confucianism, known as Neo-Confucianism, was adopted as official government policy. In addition to teaching proper behavior, Neo-Confucian scholars and officials discussed spiritual questions like what made human beings do bad things even if their basic nature was good.

> Conduct some research to find the title usually given in English to a book containing Confucius's ideas. Write that title here.
>
> _______________________
>
> _______________________

> Before the Song dynasty, what religious belief probably had a negative effect on the popularity of Confucianism in China?
>
> _______________________
>
> _______________________

Section 4, *continued*

SCHOLAR-OFFICIALS

The Song dynasty took another major step that would affect the Chinese imperial state for centuries to come. The Song improved the system by which people went to work for the government. These workers formed a large **bureaucracy** by passing a series of written **civil service** examinations.

The tests covered both the traditional teachings of Confucius and related ideas. Because the tests were extremely difficult, students spent years preparing for them. Candidates had a strong incentive for studying hard. Passing the tests meant life as a **scholar-official**, whose benefits included considerable respect and reduced penalties for breaking the law.

The civil service examination system helped ensure that talented, intelligent people became scholar-officials. This system was a major factor in the stability of the Song government.

Name a well-known government division today that has a large bureaucracy.

____

____

Draw a picture of what you think a Song scholar-official might look like.

CHALLENGE ACTIVITY

Critical Thinking: Drawing Inferences Write a short essay on the relation between the Song dynasty development of civil service and the Confucian ideals of *ren* and *li*.

Interactive Reader and Study Guide

History of Ancient China

Section 5

MAIN IDEAS

1. The Mongol Empire included China, and the Mongols ruled China as the Yuan dynasty.
2. The Ming dynasty was a time of stability and prosperity.
3. The Ming brought great changes in government and relations with other countries.

Key Terms and Places

Beijing present-day city near the capital of the Yuan dynasty

Forbidden City a huge palace complex that included hundreds of imperial residences, temples, and other government buildings

isolationism a policy of avoiding contact with other countries

Section Summary

THE MONGOL EMPIRE

In 1206, a powerful Mongol leader known as Genghis Khan (jeng-uhs KAHN) led huge armies through much of Asia and Eastern Europe. He first led his armies into northern China in 1211, then headed south. By the time of Genghis Khan's death in 1227, all of northern China was under Mongol control.

Genghis Khan's grandson, Kublai Khan (KOO-bluh KAHN), completed the conquest of China and declared himself emperor in 1279. This began the Yuan dynasty, a period also known as the Mongol Ascendancy.

Kublai Khan did not force the Chinese to accept Mongol customs, but he did try to control them. One way was by having the Chinese pay heavy taxes, which were used to pay for building projects. One such project was the building of a new capital, Dadu, near the present-day city of **Beijing**.

Kublai Khan's regime preserved much of the structure of the Song dynasty, including the civil service and trade routes. The Italian merchant Marco Polo, who traveled in China between

> **How many years did it take for the Mongol armies to conquer all of China?**
>
> ______________________
>
> ______________________

> **Which two aspects of Song civilization would you say Kublai Khan appreciated the most?**
>
> ______________________
>
> ______________________

Section 5, *continued*

1271 and 1295, wrote of his travels and sparked
Europeans' interest in China.

Two failed campaigns against Japan and
expensive public works projects weakened the Yuan
dynasty. Many Chinese groups rebelled, and in 1368,
Zhu Yuanzhang (JOO yoo-ahn-JAHNG) took control
and founded the Ming dynasty.

THE MING DYNASTY

The Ming dynasty lasted nearly 300 years, from
1368 to 1644. Ming China proved to be one of the
most stable and prosperous times in Chinese
history. Great Ming achievements include the
remarkable ships and voyages of Zheng He (juhng
HUH), the Great Wall of China, and the **Forbidden
City** in Beijing. The Forbidden City was a massive
palace of residences, temples, and government
buildings. Common people were not allowed to
enter the Forbidden City.

> **What do you think was the original reason for building the Great Wall?**
>
> _______________________
> _______________________
> _______________________

CHINA UNDER THE MING

Emperors during the Ming dynasty worked to
eliminate foreign influences from Chinese society.
China entered a period of **isolationism**. Ironically,
the consequences of this policy included a weakness
that allowed opportunistic Westerners to seize
considerable power in some parts of China as
China's imperial glory faded.

> **Name another major country whose history includes a period of isolationism.**
>
> _______________________
> _______________________

CHALLENGE ACTIVITY

Critical Thinking: Drawing Inferences Draw a street map of an
imaginary city. Include a "forbidden city" within it that is restricted to
a certain group of your choosing.

  Interactive Reader and Study Guide

The Indian Subcontinent

CHAPTER SUMMARY

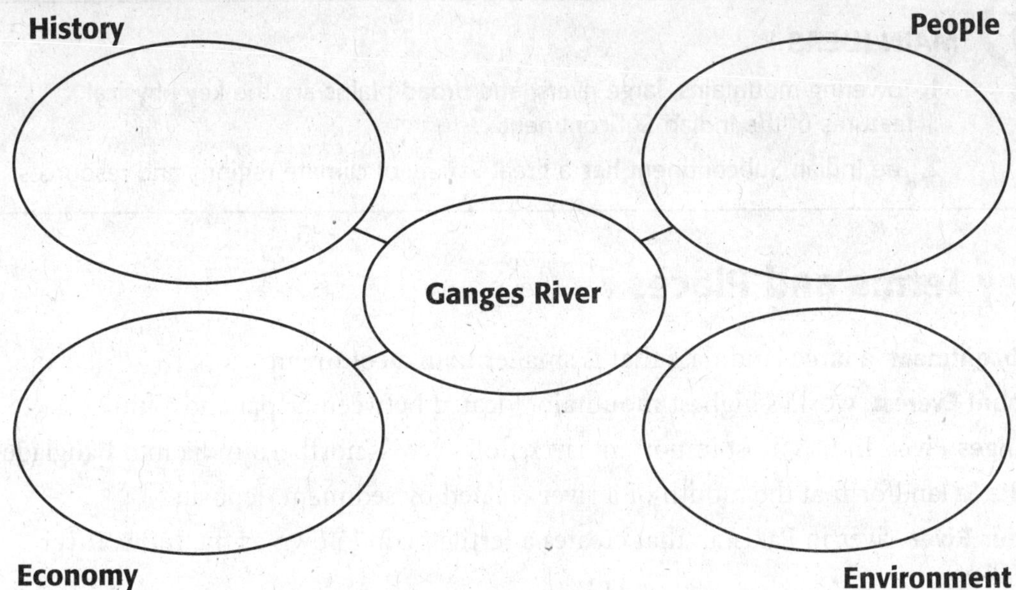

COMPREHENSION AND CRITICAL THINKING

Use the graphic organizer to answer the following questions.

1. **Evaluate** In what ways has the Ganges River played a part in the history, people, economy, and environment of the Indian Subcontinent? Choose from the following terms, and write your answer in the spaces in the graphic organizer. (Note: Some answers are not used while others are used more than once.)

   Harappan civilization, causes floods, location of major cities, Aryan civilization, rich soil deposits, Diwali festival, international trade, agricultural output, Gupta Empire, keeps foreigners out, population growth, causes monsoons

2. **Make Inferences** Is the Indus River important to Pakistan in the same ways that the Ganges River is important to India? Give reasons for your answer.

   ____

   ____

3. **Predict** Will the Ganges River play an important part in India's future? Why or why not?

   ____

   ____

Section 1

MAIN IDEAS

1. Towering mountains, large rivers, and broad plains are the key physical features of the Indian Subcontinent.

2. The Indian Subcontinent has a great variety of climate regions and resources.

Key Terms and Places

subcontinent a large landmass that is smaller than a continent

Mount Everest world's highest mountain, located between Nepal and China

Ganges River India's most important river, flows across northern India into Bangladesh

delta a landform at the mouth of a river created by sediment deposits

Indus River river in Pakistan that creates a fertile plain known as the Indus River Valley

monsoons seasonal winds that bring either moist or dry air to an area

Section Summary

PHYSICAL FEATURES

The Indian Subcontinent is made up of the countries Bangladesh, Bhutan, India, Maldives, Nepal, Pakistan, and Sri Lanka. This subcontinent is also known as South Asia. A **subcontinent** is a large landmass that is smaller than a continent. Huge mountains separate the Indian Subcontinent from the rest of Asia—the Hindu Kush in the northwest and the Himalayas along the north. Lower mountains, called the Ghats, run along India's eastern and western coasts. The Himalayas stretch about 1,500 miles across and are the highest mountains in the world. The highest peak, **Mount Everest**, rises 29,035 feet (8,850 m) above sea level. Pakistan's K2 is the world's second tallest peak. Two major river systems originate in the Himalayas. They have flooded the surrounding land, creating fertile plains. The **Ganges River** flows across northern India. The area along the Ganges is called the Ganges Plain. It is India's farming heartland. In Bangladesh the Ganges River joins other rivers to

> Circle the names of the seven countries in South Asia.

> Underline the world's two highest mountain peaks.

                              Interactive Reader and Study Guide

Section 1, *continued*

form a huge **delta**, a landform created by sediment deposits. Pakistan's **Indus River** also forms a fertile plain, the Indus River Valley. This region was once home to the earliest Indian civilizations. Now, it is the most heavily populated area in Pakistan.

Other features include a hilly plateau south of the Ganges Plain called the Deccan. East of the Indus Valley is the Thar, or Great Indian Desert. In southern Nepal, the Tarai region is known for its fertile farmland and tropical jungles.

> Which river forms a fertile plain in Pakistan?
> ________________________________

CLIMATES AND RESOURCES

Nepal and Bhutan, located in the Himalayas, have a highland climate which brings cool temperatures. In the plains south of the Himalayas, the climate is humid subtropical. The rest of the subcontinent has mainly tropical climates. Central India and Sri Lanka have a tropical savanna climate, with warm temperatures year round. Bangladesh, Sri Lanka, Maldives, and parts of southwest India have a humid tropical climate, with warm temperatures and heavy rains. Southern and western India and most of Pakistan have desert and steppe climates. **Monsoons**—winds that bring either dry or moist air—greatly affect the subcontinent's climate. From June to October, summer monsoons from the Indian Ocean bring heavy rains. In winter, monsoons change direction and bring in dry air from the north.

> Underline the type of climate found in Nepal and Bhutan.

> Define monsoon in your own words on the lines below.
> ________________________________
> ________________________________

The subcontinent's fertile soil is a vital resource for the region. It allows farmers to produce tea, rice, nuts, and jute. Other important resources are timber, livestock, iron ore, coal, natural gas, and gemstones.

> Circle the resources of the Indian Subcontinent.

CHALLENGE ACTIVITY

Critical Thinking: Making Inferences Write a paragraph explaining why you think Indian civilization began in the Indus River Valley.

Interactive Reader and Study Guide

The Indian Subcontinent

Section 2

MAIN IDEAS

1. Advanced civilizations and powerful empires shaped the early history of India.
2. Powerful empires controlled India for hundreds of years.
3. Independence from Great Britain led to the division of India into several countries.
4. Religion and the caste system are two important parts of Indian culture.

Key Terms and Places

Delhi site of former Muslim kingdom in northern India

colony territory inhabited and controlled by people from a foreign land

partition division

Hinduism one of the world's oldest religions, the dominant religion of India

Buddhism religion based on the teaching of Siddhartha Gautama, the Buddha

caste system divides Indian society into groups based on birth or occupation

Section Summary

EARLY CIVILIZATIONS AND EMPIRES

The Indian Subcontinent's first urban civilization—the Harappan—was in the Indus Valley. Later the Aryans moved from Central Asia into the Indian Subcontinent. They settled along the Indus and Ganges rivers. An Aryan language, Sanskrit, became the basis for many languages, including Hindi. Aryan customs mixed with those of other people, forming much of India's culture.

> Circle the Aryan language that is the basis for Hindi.

The Mauryan people then conquered most of the subcontinent by about 320 BC. However, after the death of Asoka, one of the greatest Mauryan rulers, the empire split up. In the AD 300s the Gupta Empire united most of northern India. Trade and culture thrived under Gupta rulers until around 550.

> Underline the name of one of the greatest Mauryan emperors.

POWERFUL EMPIRES

In the late 600s Muslim armies invaded India. They set up a kingdom at **Delhi** in northern India and formed the Mughal Empire. Trade and culture

    Interactive Reader and Study Guide

Section 2, *continued*

flourished during this period, especially under Akbar, one of India's greatest rulers.

In the 1600s, as the Mughal Empire declined, the British East India Company gained valuable trading rights. India eventually became a British **colony**, a territory controlled by people from a foreign land. Many Indians rebelled against the British company. The British government crushed the rebellion and began ruling India directly.

> Underline the sentence that explains what a colony is.

INDEPENDENCE AND DIVISION

By the late 1800s many Indians were upset at British rule. One group formed the Indian National Congress to gain more rights. Later, Mohandas Gandhi led nonviolent protests to gain Indian independence. However, Muslims feared having little say in the new India. To avoid civil war, the British agreed to the **partition**, or division of India. As a result, two independent countries, India and Pakistan, were formed in 1947. Later Sri Lanka and Maldives gained independence. Bangladesh broke away from Pakistan to form a new nation in 1971.

> Why was India divided into two independent countries?
> ____________________
> ____________________

INDIAN CULTURE

India is the birthplace of **Hinduism**, one of the world's oldest religions. Hindus believe everything in the universe is part of one spirit called Brahman. The goal of Hindus is to reunite their souls with Brahman. Another religion, **Buddhism** is based on the teachings of the Buddha, Siddhartha Gautama. It also began in India. The Buddha taught that people can achieve a state of peace called nirvana.

> Circle the two religions that began in India.

Indian society was organized into a **caste system**, which divides people into different classes. A person's class was based on his or her birth or occupation.

> Explain the caste system in your own words on the lines below.
> ____________________
> ____________________

CHALLENGE ACTIVITY

Critical Thinking: Comparing and Contrasting Write a paragraph that explains how Mughal and British rule were alike and different.

The Indian Subcontinent

Section 3

MAIN IDEAS

1. Daily life in India is centered around cities, villages, and religion.
2. Today India faces many challenges, including a growing population and economic development.

Key Terms and Places

Mumbai (Bombay) one of India's largest cities

Kolkata (Calcutta) one of India's largest cities

urbanization increase in the percentage of people who live in cities

green revolution program that encouraged farmers to adopt modern agricultural methods

Section Summary

DAILY LIFE IN INDIA

Today more than one billion people live in India. India has many different ethnic groups, religions, and lifestyles. City life, village life, and religion help unite India's people.

India's two largest cities, **Mumbai** (**Bombay**) and **Kolkata** (**Calcutta**), are among the world's most populous. Cities such as Mumbai and Bangalore have universities, research centers, and high-tech businesses. But most city-dwellers have a hard time making a living. Many live in shacks, with no plumbing and little clean water.

More than 70 percent of India's population lives in villages. Most work on farms and live with extended families in simple homes. In many areas paved roads and electricity have only recently been provided.

Religion plays a key role in Indian villages and cities. Most people practice Hinduism, while others practice Islam, Buddhism, Jainism and Sikhism. Diwali, the festival of lights, is one of India's most popular religious events. Diwali celebrates Hindu, Sikh, and Jain beliefs.

> Circle the number of people who live in India.

> Underline the sentence that describes the living conditions of many city-dwellers.

> Circle the names of religions practiced in India.

   Interactive Reader and Study Guide

Section 3, *continued*

INDIA'S CHALLENGES

Only China has more people than India. India's population has doubled since 1947. This great increase in population has strained India's environment and resources such as food, housing, and schools. **Urbanization**—an increase in the percentage of a country's people who live in cities—has greatly affected India's cities. In search of jobs, millions of people moved to India's cities.

Since gaining independence, India's leaders have improved its government and economy. India is now one of the strongest nations in Asia—and the world's largest democracy. India faces two main challenges. The first is how to provide resources for its growing population. The second is how to resolve conflicts with its neighbor, Pakistan.

India's gross domestic product (GDP) makes it one of the world's top five industrial nations. However, GDP per capita (per person) is only $3,100, so millions of Indians live in poverty. To help address this problem, India's government began a program called the **green revolution**, which encouraged farmers to use more modern farming methods. Recently, the government has worked to bring high-tech businesses to India.

CHALLENGE ACTIVITY

Critical Thinking: Making Inferences If India is one of the world's top five industrial nations, why do so many of its people live in poverty? Write a short paragraph to explain your answer.

 Interactive Reader and Study Guide

The Indian Subcontinent

Section 4

MAIN IDEAS

1. Many different ethnic groups and religions influence the culture of India's neighbors.
2. Rapid population growth, ethnic conflicts, and environmental threats are major challenges to the region today.

Key Terms and Places

Sherpas ethnic group from the mountains of Nepal

Kashmir a region which both India and Pakistan claim control over

Dhaka capital of Bangladesh and its largest city

Kathmandu capital of Nepal and its largest city

Section Summary

CULTURE

India's neighbors have different ways of life. Their cultures reflect the customs of many ethnic groups. For example, the **Sherpas** in Nepal often serve as guides through the Himalayas. Many of the Tamil in Sri Lanka came from India to work on plantations.

People of the region also have different religious beliefs. Like India, most of its neighbors have one major religion. For example, most people in Pakistan and Bangladesh practice Islam. Hinduism is the major religion in Nepal while Buddhism is the major religion in Sri Lanka and Bhutan.

> Circle the countries where most people practice Islam.

THE REGION TODAY

Since its creation in 1947, Pakistan has not had a stable government. Rebellions and assassinations have hurt the country. Pakistan also faces the challenges of overpopulation and poverty. These challenges could cause even more instability. Pakistan has also clashed with India over control of the territory of **Kashmir**. Pakistan controls western Kashmir and India controls the east. However, both countries claim control over this region.

> What challenges does Pakistan face?
>
> _______________________
>
> _______________________

Section 4, *continued*

Since 2001 Pakistan has helped the United States fight terrorism. Many people, though, think terrorists still remain in Pakistan.

Bangladesh is a small country, but one of the world's most densely populated. It has about 2,734 people per square mile (1,055 per square km). More than 11.5 million people live in Bangladesh's capital, **Dhaka**. One of the country's main challenges is flooding from rivers and monsoons, which often causes heavy damage. For example, one flood left over 25 million people homeless.

> Underline the sentence that tells one of Bangladesh's main challenges.

Nepal's population is growing rapidly. Its largest city and capital, **Kathmandu**, is poor and overcrowded. Nepal also faces environmental threats. Land cleared to grow food causes deforestation, leading to soil erosion and harming wildlife. Tourists also harm its environment by leaving trash behind and using valuable resources.

> Who is responsible for some of the damage to Nepal's environment?
>
> ___________________________

Bhutan is a small, isolated mountain kingdom between India and China. After years of isolation, Bhutan formed ties with Great Britain and India in the 1900s. Bhutan has begun to modernize, building new roads, schools, and hospitals. Most of its people are farmers, growing rice, potatoes, and corn. To protect its environment and way of life, Bhutan limits the number of tourists who may visit.

> Underline the crops Bhutan's farmers grow.

Sri Lanka has been greatly influenced by its close neighbor, India. Two of Sri Lanka's main ethnic groups—the Tamil and the Sinhalese—have Indian roots. Unfortunately, these two groups are often in conflict with one another. Despite a cease-fire in 2002, violence between them goes on. In 2004, an Indian Ocean tsunami struck Sri Lanka, killing thousands. More than 500,000 were left homeless. Sri Lanka is still trying to rebuild its fishing and agricultural industries.

> Circle the two main ethnic groups in Sri Lanka.

CHALLENGE ACTIVITY

Critical Thinking: Making Predictions How can people in Sri Lanka rebuild homes and businesses destroyed by the tsunami in 2004?

China, Mongolia, and Taiwan

CHAPTER SUMMARY

	China	**Mongolia**	**Taiwan**
Geography	plains, mountains, deserts, plateaus	desert, grasslands, mountains	plains, mountains
History	ancient dynasties Communist revolution	Mongol Empire Chinese control Soviet influence	controlled by China, Japan, Chinese Nationalists
Culture	Buddhism, Daoism, Confucianism	traditional herders	Chinese

COMPREHENSION AND CRITICAL THINKING

Use information from the graphic organizer to answer the following questions.

1. Describe What geographical feature(s) does China share with Mongolia and Taiwan?

____

____

2. Interpret Write the words *tropical* and *nomadic* in the appropriate spaces in the chart.

3. Elaborate Which cultural details could be added to China and Taiwan?

____

____

China, Mongolia, and Taiwan

Section 1

MAIN IDEAS

1. Physical features of China, Mongolia, and Taiwan include mountains, plateaus and basins, plains, and rivers.
2. China, Mongolia, and Taiwan have a range of climates and natural resources.

Key Terms and Places

Himalayas the world's tallest mountain range

Plateau of Tibet the world's highest plateau, located in southwest China

Gobi located in Mongolia, the world's coldest desert

North China Plain fertile plain in east China

Huang He the Yellow River, a river in northern China that often floods

loess fertile, yellowish soil

Chang Jiang the Yangzi River, Asia's longest river, flows across central China

Section Summary

PHYSICAL FEATURES

China has a range of physical features. These include the world's tallest mountains, as well as some of the world's driest deserts and longest rivers. Mongolia and Taiwan are two of China's neighbors. Mongolia is a dry, landlocked country. Taiwan is a green tropical island.

> How are Mongolia and Taiwan different?
>
> _______________________
>
> _______________________

Mountains are found in much of the region. The **Himalayas** run along the border of southwest China. They are the highest mountains in the world. The highest plateau in the world—the **Plateau of Tibet**—is also located in southwest China. Many of the region's mountain ranges are separated by plateaus, basins, and deserts.

> Where in China are the Himalayas and the Plateau of Tibet located?
>
> _______________________
>
> _______________________

The Taklimakan Desert is located in western China. Swirling sandstorms are frequent here. Another desert, the **Gobi**, is located in Mongolia. It is the world's coldest desert.

     Interactive Reader and Study Guide

Section 1, *continued*

Most Chinese live in the eastern **North China Plain**. This region is made up of low plains and river valleys. In Taiwan most people live on a plain on the west coast.

> Underline the region where most Chinese live.

Two long rivers run west to east across China. One of these, the **Huang He**, or the Yellow River, picks up a yellowish, fertile soil called **loess**. When the river floods, it deposits the loess, enriching the farmland along the banks. But many people are killed by these floods. Another river, the **Chang Jiang**, or the Yangzi River, flows across central China. It is Asia's longest river and a major transportation route.

> How does the Huang He both help and harm people?
>
> _____________________
>
> _____________________

CLIMATE AND RESOURCES

Climate varies widely across the region. The tropical southeast is warm to hot. There, monsoons bring heavy rains in the summer. Violent storms called typhoons bring heavy winds and rain in the summer and fall. The climate in the north and west is mainly dry. Temperatures across this area vary. The climate in the northeast is quite different. It is drier and colder. In the winter, temperatures can drop below 0°F (-18°C).

> What is a typhoon?
>
> _____________________

The region has a variety of natural resources. Farmland is an important resource in both China and Taiwan. Taiwan grows a variety of crops, including sugarcane, tea, and bananas. China also has many mineral, metal, and forest resources. Mongolia's natural resources include minerals and livestock.

> Circle an important resource in both China and Taiwan.

CHALLENGE ACTIVITY

Critical Thinking: Making Generalizations Write a journal entry describing your travels through the region, such as hiking in the Himalayas, traveling with nomads across the Gobi desert, or visiting a city. What are your general impressions?

China, Mongolia, and Taiwan

Section 2

MAIN IDEAS

1. Family lines of emperors ruled China for most of its early history.
2. In China's modern history, revolution and civil war led to a Communist government.
3. China has the world's most people as well as a rich culture shaped by ancient traditions.

Key Terms

dynasty a series of rulers from the same family line

dialect a regional version of a language

Daoism belief system that stresses living simply and in harmony with nature

Confucianism a philosophy based on the ideas and teachings of Confucius

pagodas Buddhist temples that have multi-storied towers with an upward curving roof at each floor

Section Summary

CHINA'S EARLY HISTORY

China's civilization has lasted for about 4,000 years. For most of its long history, China was ruled by dynasties. A **dynasty** is a series of rulers from the same family line.

> Underline the definition of *dynasty*.

China limited contact with the outside world for much of its history. But many people wanted Chinese goods, such as silk and tea. Some European powers forced China to open up trade in the 1800s.

CHINA'S MODERN HISTORY

Rebels forced out China's last emperor in 1911. Over time, the Nationalists and the Communists fought a long civil war. The Communists won in 1949, and the Nationalists fled to Taiwan.

> Who won China's civil war? When did the war end?
>
> ____________________
>
> ____________________

Mao Zedong led the new Communist government in China. Some people's lives improved. But people who criticized the government were punished, and freedoms were limited. Some economic programs failed, causing famine and other problems.

  Interactive Reader and Study Guide

Section 2, *continued*

Mao died in 1976. China's next leader was Deng Xiaoping. Deng worked to modernize and improve the economy. The economy began growing rapidly. Later leaders continued economic reforms.

> How were the changes made by Mao Zedong different from the changes made by Deng Xiaoping?
> _______________________
> _______________________

CHINA'S PEOPLE AND CULTURE

China has 1.3 billion people. Most are crowded in the East, on the Manchurian and North China plains. The majority belong to the Han ethnic group. Many speak Mandarin, China's official language. Others speak a **dialect**, a regional version of a language.

> Circle the name of China's official language.

Modern China is influenced by many values and beliefs. The main belief systems are Buddhism and Daoism. **Daoism** stresses living simply and in harmony with nature. Buddhists believe moral behavior, kindness, and meditation can lead to peace.

Many Chinese blend elements of these religions with **Confucianism**. This philosophy is based on the ideas and teachings of Confucius. It stresses family, moral values, and respect for one's elders.

> What is Confucianism?
> _______________________
> _______________________

The Chinese have rich artistic traditions. Crafts are made from materials such as bronze, jade, ivory, silk, wood, and porcelain. Paintings on silk paper, calligraphy, poetry, and opera are also common. Traditional architecture features wooden buildings with upward-curving roofs. Buddhist temples called **pagodas** have multi-storied towers with roofs that curve upward at each floor. Popular culture includes sports such as martial arts and activities such as visiting karaoke clubs.

CHALLENGE ACTIVITY

Critical Thinking: Comparing and Contrasting Imagine that you are a Chinese emperor and that Europeans want to begin trading with China. Would you allow such contact with the outside world? Why or why not? Explain your answer in a brief essay.

China, Mongolia, and Taiwan

Section 3

MAIN IDEAS

1. China's booming economy is based on agriculture, but industry is growing rapidly.
2. China's government controls many aspects of life and limits political freedom.
3. China is mainly rural, but urban areas are growing.
4. China's environment faces a number of serious problems.

Key Terms and Places

command economy an economic system in which the government owns all businesses and makes all decisions

Beijing China's capital

Tibet the Buddhist region in southwest China

Shanghai China's largest city

Hong Kong important center of trade and tourism in southern China

Section Summary

CHINA'S ECONOMY

Communist China used to have a **command economy**, in which the government owns all the businesses and makes all decisions. Because it had major economic problems, China began allowing some aspects of a market economy in the 1970s. In a market economy people can make their own decisions and keep the profits they earn.

China's mixed economic approach has helped its economy boom. Today it is the world's second-largest economy.

More than half of all Chinese workers are farmers. They are able to produce a lot of food. However, industry and manufacturing are the most profitable part of China's economy. Economic growth has improved wages and living standards in China. Still, many rural Chinese remain poor. Many do not have work.

How is a command economy different from a market economy?

____

Underline the reason why China is able to grow a lot of food.

CHINA'S GOVERNMENT

China's citizens have little political freedom. In 1989, a huge protest took place in China's capital, **Beijing**. About 100,000 people gathered in Tiananmen Square to demand more political rights and freedoms. The government crushed the protest.

China has also put down ethnic rebellions. One revolt took place in 1959 in **Tibet**, a Buddhist region. China has since restricted Tibetans' rights.

Other nations have criticized these actions. They have considered stopping or limiting trade with China until it shows more respect for human rights.

> How have some countries responded to China's actions against human rights?
>
> ____________________________

RURAL AND URBAN CHINA

Most of China's people live in small rural villages. But many people are moving to China's growing cities. The country's largest city is **Shanghai**. China's second-largest city is Beijing. Beijing is an important cultural and political center. Two important cities on the southern coast are **Hong Kong** and Macau. These were once under European control but have recently been returned to China.

> Circle the names of two of China's cities that were under European control until recently.

CHINA'S ENVIRONMENT

China's growth has created environmental problems. These include air and water pollution. Forestland and farmland have been lost. China is working to solve these problems.

CHALLENGE ACTIVITY

Critical Thinking: Drawing Conclusions Imagine that you are living in China and that you have a new pen pal who is curious about what it is like to live in your country. Write a brief letter describing life in China.

China, Mongolia, and Taiwan

Section 4

MAIN IDEAS

1. Mongolia is a sparsely populated country where many people live as nomads.
2. Taiwan is a small island with a dense population and a highly industrialized economy.

Key Terms and Places

gers large, circular, felt tents that are easy to put up, take down, and move

Ulaanbaatar Mongolia's capital and only large city

Taipei Taiwan's capital and main financial center

Kao-hsiung Taiwan's main seaport and a center of heavy industry

Section Summary

MONGOLIA

The people of Mongolia have a proud and fascinating history. In the late 1200s, Mongolia was perhaps the greatest power in the world. The Mongol Empire stretched from Europe to the Pacific Ocean. Over time, the empire declined. China conquered Mongolia in the late 1600s.

Mongolia declared independence from China in 1911. The Communists gained control 13 years later. The Soviet Union was a strong influence in Mongolia. But in the early 1990s, the Soviet Union collapsed. Since then, Mongolians have worked for democracy and a free-market economy.

Many Mongolians have a traditional way of life. Nearly half live as nomads, herding livestock. Many live in **gers**. These large, circular, felt tents are easy to put up, take down, and move. Horses play an important role in the lives of the nomads.

Mongolia has a small population. Only about 2.7 million people live in the entire country. Mongolia's capital and only large city is **Ulaanbaatar**. One in four Mongolians lives there. The country's main industries include textiles, carpets, coal, copper, and oil. Mongolia produces

Circle the name of the country that had a strong influence in Mongolia for much of the 1900s.

How do most Mongolians live?

____

Circle the name of Mongolia's capital city.

livestock but very little other food. The country faces both food and water shortages.

TAIWAN

Both China and Japan controlled Taiwan at different times. In 1949 the Chinese Nationalists took over the island. They had left the Chinese mainland after the Communists took over. The Nationalists ruled Taiwan under martial law, or military rule, for 38 years. Today, Taiwan's government is a multiparty democracy.

Why did Chinese Nationalists come to Taiwan?

____

____

Tension remains between China and Taiwan. China claims that Taiwan is a rebel part of China. Taiwan claims to be China's true government.

Taiwan's history is reflected in its culture. Most Taiwanese are descendants of people from China. As a result, Chinese ways are an important part of Taiwan's culture. Japan's influence is also seen. European and American practices and customs are beginning to be seen in Taiwan's cities.

What are some cultural influences on Taiwanese culture? Which is the major influence?

____

Taiwan is a modern country with about 23 million people. Most Taiwanese live in cities on the island's western coastal plain. The rest of the country is mountainous. The two largest cities are **Taipei** and **Kao-hsiung**. Taipei is Taiwan's capital. It is a main financial center. Kao-hsiung is a center of heavy industry and Taiwan's main seaport.

Underline the place where most Taiwanese live. What is the rest of the country like?

____

____

CHALLENGE ACTIVITY

Critical Thinking: Comparing and Contrasting Imagine that you are a government official from Taiwan visiting Mongolia. Write a report explaining why Mongolia and Taiwan should or should not become trade partners.

Japan and the Koreas

CHAPTER SUMMARY

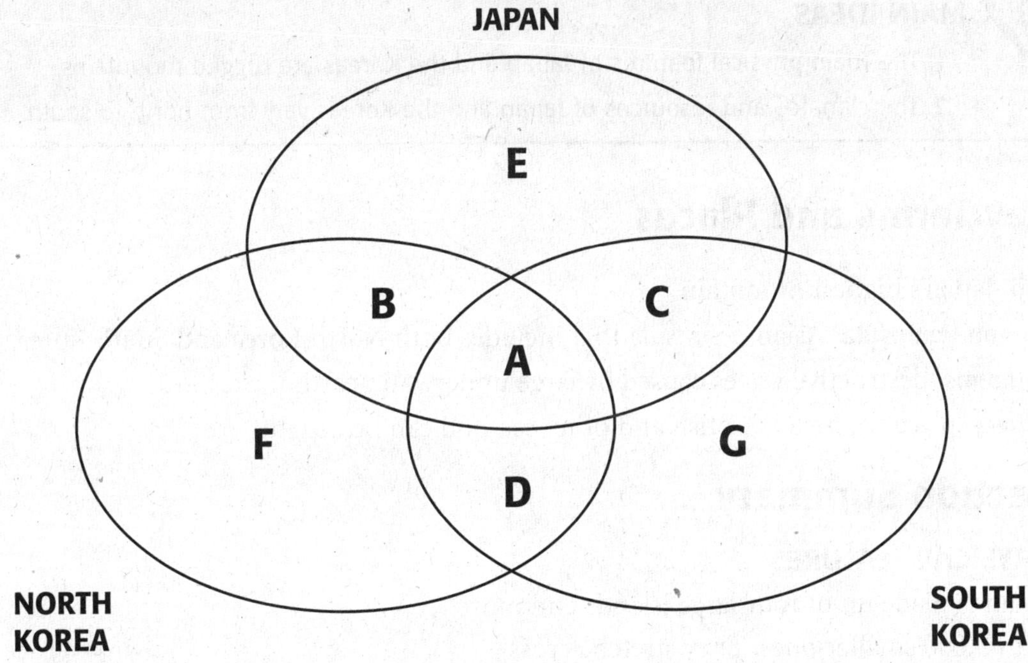

COMPREHENSION AND CRITICAL THINKING

For each item, write the letter of the correct area in the graphic organizer.

______ **1.** rich in mineral resources

______ **2.** early history influenced by China

______ **3.** Westernized nation with strong manufacturing economy

______ **4.** majority of land is covered by rugged mountains

______ **5.** religious practice combines Shintoist and Buddhist traditions

6. Sequence Write a short passage describing the history of Japan and the Korean Peninsula, including at least five major periods or dates.

____

____

____

____

____

____

Japan and the Koreas

Section 1

MAIN IDEAS

1. The main physical features of Japan and the Koreas are rugged mountains.
2. The climates and resources of Japan and the Koreas vary from north to south.

Key Terms and Places

Fuji Japan's highest mountain

Korean Peninsula Asian peninsula that includes both North Korea and South Korea

tsunamis destructive waves caused by large underwater earthquakes

fishery place where lots of fish and other seafood can be caught

Section Summary

PHYSICAL FEATURES

Japan is made up of four large islands and more than 3,000 smaller ones. They stretch across 1,500 hundred miles of ocean, about the length of the Eastern United States coastline. But they include only about as much land area as California. Most people live on the four largest islands, which are Hokkaido, Honshu, Shikoku, and Kyushu. Mountains cover about 75 percent of Japan. The Japanese Alps are Japan's largest mountain range. Japan's highest mountain, **Fuji**, is not in any mountain range, but is an isolated volcanic peak in eastern Honshu. It has become a symbol of Japan, and is considered sacred by some people. Many shrines and temples have been built around it.

Circle the names of Japan's largest islands.

The **Korean Peninsula** juts south from the Asian mainland, and is divided between North and South Korea. Rugged mountains run along the eastern coast, and plains can be found on the western coast and in the river valleys. Korea has more rivers than Japan. Most of them flow westward and empty in the Yellow Sea.

Where on the Korean Peninsula are the mountains located?

____

____

Interactive Reader and Study Guide

Section 1, *continued*

Japan is subject to volcanic eruptions, earthquakes, and **tsunamis,** which are destructive waves caused by underwater earthquakes. Korea does not have many earthquakes or volcanoes. Like Japan, Korea is subject to huge storms, called typhoons, that sweep in from the Pacific Ocean.

CLIMATE AND RESOURCES

Just as Japan and the Koreas have many similar physical features, they also have similar climates. In both places, climate varies from north to south. The northern regions have a humid continental climate with cool summers, long, cold winters, and a short growing season. In the south, a humid subtropical climate brings mild winters and as much as 80 inches of rain each year. Most of the rain falls during the hot, humid summers, which is also when typhoons occur.

Unlike the rest of the region, North Korea is rich in mineral resources such as iron and coal. Both of the Koreas use their quick-flowing rivers to generate hydroelectric power. Japan has one of the world's strongest fishing economies. The islands lie near one of the world's most productive **fisheries**, which are areas where lots of fish and seafood can be caught. Huge fishing nets are used to catch the large number of fish needed to serve Japan's busy fish markets.

CHALLENGE ACTIVITY

Critical Thinking: Analyzing Write a paragraph describing how physical features and climate affect daily life in Japan.

Section 2

MAIN IDEAS

1. The early histories of Japan and Korea were closely linked, but the countries developed very differently.

2. Japanese culture blends traditional customs with modern innovations.

3. Though they share a common culture, life is very different in North and South Korea.

Key Terms and Places

Kyoto Japan's imperial capital, known before as Heian

shoguns powerful military leaders of imperial Japan

samurai highly trained warriors

kimonos traditional Japanese robes

kimchi Korean dish made from pickled cabbage and spices

Section Summary

HISTORY

China has had much influence on both Japan and the Koreas. One example is Buddhism, which was once the main religion in both countries. Japan's first government was modeled after China's. Japan's emperors made their capital Heian (now called **Kyoto**) a center for the arts. Some of them paid so much attention to the arts that they allowed generals called **shoguns** to take control. Shoguns had armies of warriors called **samurai** who helped them rule Japan until 1868, when a group of samurai gave power back to the emperor. During World War II, the Japanese brought the United States into the war by bombing its naval base at Pearl Harbor. To end the war, the Americans dropped atomic bombs on the Japanese cities of Hiroshima and Nagasaki.

China ruled the Korean Peninsula for centuries. Later, the Japanese invaded. After World War II, Korea became independent. However, the Soviet Union helped Communists take control in the north while the United States helped to form a

> Underline the sentence that explains how the shoguns gained power.

> Which country helped Communists take control in North Korea?
>
> _______________________

democratic government in the south. In 1950 North Korea invaded the south, wanting to unite Korea. The United States and other countries helped the South Koreans remain separate.

JAPANESE CULTURE

The Japanese system of writing uses two types of characters. Some characters, called kanji, each represent a single word. Others, called kana, each stand for a part of a word. Most Japanese people combine elements of Shinto and Buddhism in their religious practices. In the Shinto tradition—native to Japan—everything in nature is believed to have a spirit, or *kami,* which protects people. Through Buddhism, people in Japan have learned to seek enlightenment and peace. Most people in Japan wear Western style clothing, but many also wear the traditional **kimono** on special occasions. Traditional forms of art include Noh and Kabuki plays.

Circle the names of the kinds of characters used to write Japanese.

What do Japanese people wear most of the time?

____

KOREAN CULTURE

People in both North and South Korea speak Korean. Unlike Japanese, written Korean uses an alphabet. In the past, most Koreans were Buddhist or Confucianist. Today, about one fourth of South Koreans are Christian. Communist North Korea discourages people from practicing any religion.

   The people of Korea have kept many old traditions, such as **kimchi**, a dish made from pickled cabbage. Traditional Korean art forms remain, especially in North Korea, where the Communists think Korean culture is the best in the world. In the south, people have adopted new ways of life and forgotten some of their traditions.

Underline the names of religions practiced in Korea.

CHALLENGE ACTIVITY

Critical Thinking: Sequencing Write a paragraph describing three periods or events in Japan's history.

Japan and the Koreas

Section 3

MAIN IDEAS

1. Since World War II, Japan has developed a democratic government and one of the world's strongest economies.
2. A shortage of open space shapes daily life in Japan.
3. Crowding, competition, and pollution are among Japan's main issues and challenges.

Key Terms and Places

Diet Japan's elected legislature

Tokyo capital of Japan

work ethic belief that work in itself is worthwhile

trade surplus exists when a country exports more goods than it imports

tariff fee a country charges for exports or imports

Osaka Japan's second largest city

Section Summary

GOVERNMENT AND ECONOMY

Although Japan's emperor is the country's official leader, he has little power and his main role is to act as a symbol. In Japan today power rests with an elected prime minister and a legislature, called the **Diet**, that govern from the capital of **Tokyo**. Starting in the 1950s, Japan has become an economic powerhouse. Japanese companies use the latest manufacturing techniques to make goods such as cars, televisions, and DVD players. The government has helped Japanese companies succeed by controlling production and planning for the future. The strong **work ethic** of Japan's workers has also helped. They are very well-trained and loyal. Most goods made in Japan are for export to places like the United States. Japan exports much more than it imports, causing a huge **trade surplus**, which has added to Japan's wealth. The amount of imported goods is kept low through high **tariffs**, or fees, which are added to their cost.

> Circle the terms that explain who has power in Japan.

> What happens to most goods manufactured in Japan?
>
> _______________________

 Interactive Reader and Study Guide

Japan's economic success is due to its manufacturing techniques, not natural resources. Japan must import most of the raw materials it uses to make goods, and much of its food.

DAILY LIFE

Japan is densely populated, and cities such as Tokyo are very crowded. Almost 30 million people live near Tokyo, making land scarce and expensive. Therefore, Tokyo has many tall buildings to get the most from scarce land. They also locate shops underground. Some hotels save space by housing guests in tiny sleeping chambers. Crowded commuter trains bring many people to work in Tokyo every day. For fun, people visit parks, museums, baseball stadiums, an indoor beach, and a ski resort filled with artificial snow.

About how many people live in the Tokyo area?

____

Underline the places people in Tokyo visit for fun.

Other cities include **Osaka**, Japan's second largest city, and Kyoto, the former capital. Japan's major cities are linked by efficient, high-speed trains, which can travel at more than 160 miles per hour.

Most people live in cities, but some live in villages or on farms. The average farm in Japan is only 2.5 acres, too small for most farmers to make a living, forcing many to look for jobs in the city.

ISSUES AND CHALLENGES

Despite its success, Japan is faced with several challenges. The lack of space due to dense population is a growing problem. In recent years, countries such as China and South Korea have taken business from Japanese companies. Pollution is another problem of growing concern.

Underline Japan's issues and challenges.

CHALLENGE ACTIVITY

Critical Thinking: Evaluating Information Write a paragraph about which of Japan's issues and challenges you think is the most serious. Support your choice with details from the text.

Japan and the Koreas

Section 4

> **MAIN IDEAS**
> 1. The people of South Korea today have freedom and economic opportunities.
> 2. The people of North Korea today have little freedom or economic opportunity.
> 3. Some people in both South and North Korea support the idea of Korean reunification.

Key Terms and Places

Seoul the capital of South Korea

demilitarized zone an empty buffer zone created to keep two countries from fighting

Pyongyang the capital of North Korea

Section Summary

SOUTH KOREA TODAY

The official name of South Korea is the Republic of Korea. It is headed by a president and assembly elected by the people. The United States helped create South Korea's government after World War II, and has helped to make its economy strong.

Like Japan, South Korea is densely populated. Its capital, **Seoul**, has some 40,000 people per square mile. Many people live near the western coastal plain, preferring it to the mountainous interior. In the cities, people live in small apartments and must cope with pollution. In the country, many South Koreans live on small farms, grow rice, beans, and cabbage, and follow traditional ways of life.

Although South Korea has a strong economy, some people are unhappy that industry has been controlled for years by only four families, and some members of these families use their power only to make themselves more wealthy. The government hopes to reform this corrupt system. South Korea is also challenged by its relationship with North Korea. Since the end of the Korean War in the 1950s, the countries have been separated by a **demilitarized zone**, an empty buffer zone patrolled by soldiers on both sides.

> Underline the sentence that explains how South Korea is corrupt.

> Underline the sentence that explains what a demilitarized zone is.

          Interactive Reader and Study Guide

NORTH KOREA TODAY

North Korea's official name, the Democratic People's Republic of Korea, is misleading. The country is a totalitarian state controlled by the Communist Party. From 1948 until 1994 it was led by the dictator Kim Il Sung. Since then his son Kim Jong Il has ruled. North Korea has a command economy in which the government makes all economic decisions. North Korea uses much of its rich mineral resources to make machinery and military supplies in out-of-date factories. Farms are run as cooperatives, but there is little good farmland and some food must be imported.

Although most North Koreans live in cities, such as the capital **Pyongyang**, their life is different from that of their neighbors to the south. Most people are poor and they are denied the rights of freedom of the press, speech, and religion.

Communist North Korea has been very isolated since the fall of the Soviet Union. Its economy has caused shortages and poverty. Also, many countries worry about North Korea's ability to make and use nuclear weapons. Negotiations are underway to resolve this issue.

KOREAN REUNIFICATION

Both North and South Korean governments have expressed support for reunification. In 2000, leaders met for the first time since the Korean War. They agreed to build a road connecting the two Koreas. However, they don't agree on the type of government the reunified country would have. South Korea prefers democracy, and North Korea prefers communism.

CHALLENGE ACTIVITY

Critical Thinking: Solving Problems Write a short paragraph describing a type of government that you think would be acceptable to people in both North and South Korea.

     Interactive Reader and Study Guide

Southeast Asia

CHAPTER SUMMARY

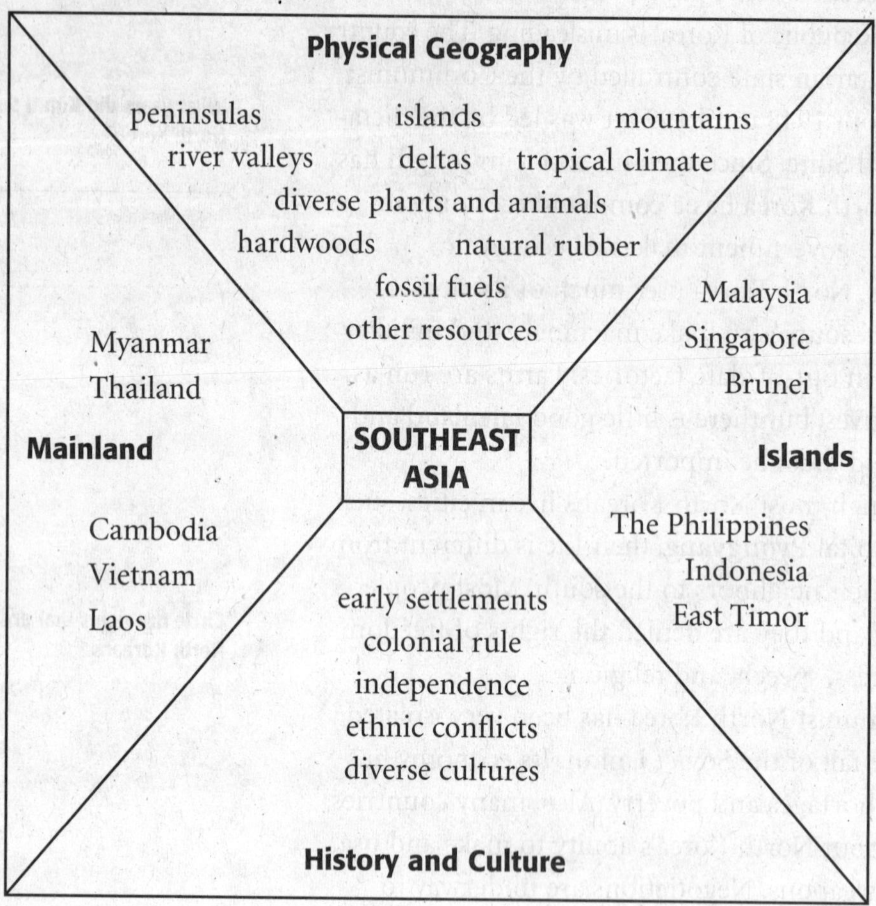

COMPREHENSION AND CRITICAL THINKING

Use information from the graphic organizer to answer the following questions.

1. Identify Which countries are neighbors on the island of Borneo?

____

2. Compare Name at least three similarities between Mainland Southeast Asia and Island Southeast Asia.

____

____

____

3. Draw a Conclusion Although Southeast Asia is rich in resources, most people are poor. Explain how this is possible.

____

____

Southeast Asia

Section 1

MAIN IDEAS

1. Southeast Asia's physical features include peninsulas, islands, rivers, and many seas, straits, and gulfs.
2. The tropical climate of Southeast Asia supports a wide range of plants and animals.
3. Southeast Asia is rich in natural resources such as wood, rubber, and fossil fuels.

Key Terms and Places

Indochina Peninsula peninsula that makes up part of Mainland Southeast Asia

Malay Peninsula peninsula that makes up part of Mainland Southeast Asia

Malay Archipelago island group that makes up part of Island Southeast Asia

archipelago a large group of islands

New Guinea Earth's second largest island

Borneo Earth's third largest island

Mekong River most important river in Southeast Asia

Section Summary

PHYSICAL FEATURES

Two peninsulas and two large island groups make up the Southeast Asia region. Mainland Southeast Asia is made up of the **Indochina Peninsula** and the **Malay Peninsula**. Island Southeast Asia is made up of the many islands of the Philippines and the **Malay Archipelago**. A large group of islands is called an **archipelago**.

> Underline the two peninsulas that make up Mainland Southeast Asia.

> Circle the two island groups that make up Island Southeast Asia.

Mainland Southeast Asia has rugged mountains, low plateaus, and river floodplains. Island Southeast Asia has more than 20,000 islands, including **New Guinea**, the world's second largest island, and **Borneo**, the world's third largest island. Island Southeast Asia is part of the Ring of Fire, where earthquakes and volcanoes often occur.

For all of Southeast Asia, water is of great importance. The region's fertile river valleys and

  Interactive Reader and Study Guide

deltas support farming and are home to many people. The **Mekong River** is the region's most important river.

CLIMATE, PLANTS, AND ANIMALS

Southeast Asia is in the tropics, the area on and around the equator. This region is generally warm all year round.

The climate on the mainland is mostly tropical savanna. Monsoon winds bring heavy rain in summer and drier air in winter there. Savannas—areas of tall grasses and some trees and shrubs—grow here.

The islands and the Malay Peninsula have a mostly humid tropical climate. Here, it's hot, muggy, and rainy all year. This climate supports tropical rain forests. These forests are home to many different plants and animals. Some animals are only found here, such as orangutans and Komodo dragons. Many plants and animals are endangered, however, due to the cutting down of the rain forest.

> What kinds of plants grow in the region's tropical savanna climate?
>
> ______________________________
>
> ______________________________

> Why are the plants and animals of the rain forest endangered?
>
> ______________________________
>
> ______________________________

NATURAL RESOURCES

Southeast Asia is rich in natural resources. Farming is very productive here thanks to the region's climate and rich soil. Rice is a major crop. Rubber tree plantations are found on Indonesia and Malaysia. The rain forests supply hardwoods and medicines. The region also has fisheries, minerals, and fossil fuels.

CHALLENGE ACTIVITY

Critical Thinking: Drawing Inferences Write an essay explaining the advantages and disadvantages of Southeast Asia's water resources for its people.

Southeast Asia

Section 2

MAIN IDEAS

1. Southeast Asia's early history includes empires, colonial rule, and independence.
2. The modern history of Southeast Asia involves struggles with war and communism.
3. Southeast Asia's culture reflects its Chinese, Indian, and European heritage.

Key Terms and Places

Timor small island that Portugal kept control of after Dutch traders drove them out of the rest of the region

domino theory idea that if one country fell to communism, other countries nearby would follow like falling dominoes

wats Buddhist temples that also serve as monasteries

Section Summary

EARLY HISTORY

Many people from China and India settled in this region. The most advanced early civilization was the Khmer Empire in what is now Cambodia. The Khmer built a huge temple, Angkor Wat. This temple showed their advanced civilization and Hindu religion. Later, the Thai settled in the Khmer area. Buddhism began to replace Hinduism in the region.

In the 1500s European countries began to set up colonies in this region. Led by Portugal, they came to colonize, trade, and spread their religious faith. For example, Spain claimed the Philippines and spread Roman Catholicism there. Later, the Dutch drove out Portugal from much of the region. **Timor**, a small island, was all that stayed under Portugal's control.

In the 1800s the British and French set up colonies and spread Christianity. The United States came into the region in 1898, when it won the Philippines from Spain in the Spanish-American War. Colonial powers ruled all of the area, except for Siam (now Thailand) by the early 1900s.

> Why did European countries come to this region?
>
> _______________________________
>
> _______________________________

> Circle the year the United States entered the region.

Interactive Reader and Study Guide

Section 2, *continued*

After World War II, the United States granted the Philippines independence. Other people in the region started to fight for freedom, too. The French left in 1954 after a bloody war in Indochina. The independent countries of Cambodia, Laos, and Vietnam were formed from this area. By 1970, most of Southeast Asia was free from colonial rule.

> **Name the three countries formed out of Indochina.**
>
> ___________________________
>
> ___________________________
>
> ___________________________

MODERN HISTORY

In Vietnam, the fighting against the French divided the country into North and South Vietnam. In South Vietnam, a civil war started. In the 1960s, the U.S. decided to send troops to South Vietnam based on the **domino theory**—the idea that if one country fell to communism, other nearby countries would fall, too. After years of fighting, North and South Vietnam became one Communist country.

> **Restate the domino theory in your own words.**
>
> ___________________________
>
> ___________________________
>
> ___________________________

In Laos and Cambodia civil wars broke out, too. Fighting lasted in Cambodia until the mid-1990s, when the United Nations helped restore peace.

CULTURE

The many different ethnic groups in this region mean that many different languages are spoken here. The main religions are Buddhism, Christianity, Hinduism, and Islam. Many grand Buddhist temples, or **wats**, which also serve as monasteries, are found in Southeast Asia. Traditonal customs are still popular, especially in rural areas. Also, many people still wear traditional clothing such as sarongs, strips of cloth worn wrapped around the body.

CHALLENGE ACTIVITY

Critical Thinking: Solving Problems Imagine you are the top government official in South Vietnam before its civil war started. Prepare and deliver a speech that tells what you will do to try to prevent this war from starting.

Southeast Asia

Section 3

MAIN IDEAS

1. The area today is largely rural and agricultural, but cities are growing rapidly.
2. Myanmar is poor with a harsh military government, while Thailand is a democracy with a strong economy.
3. The countries of Indochina are poor and struggling to rebuild after years of war.

Key Terms and Places

Yangon Myanmar's capital and major seaport

human rights rights that all people deserve, such as rights to equality and justice

Bangkok capital and largest city of Thailand

klongs canals

Phnom Penh Cambodia's capital and chief city

Hanoi capital of Vietnam, located in the north

Section Summary

THE AREA TODAY

Mainland Southeast Asia includes the countries of Myanmar, Thailand, Cambodia, Laos, and Vietnam. Because of war, harsh governments, and other problems, progress has slowed in much of this area. However, in an effort to promote political, economic, and social cooperation throughout the region, most of the countries of Southeast Asia have joined the Association of Southeast Asian Nations.

Most people in the region are farmers, living in small villages and growing rice. This area also has several big cities. They are growing rapidly as people move to them in search of work. Rapid growth, however, has led to crowding and pollution.

> Circle the countries of Mainland Southeast Asia.

> What are the effects of the rapid growth on the area's cities?
>
> _______________________
>
> _______________________

MYANMAR AND THAILAND

Although Myanmar, also called Burma, has many resources, it is a poor country. **Yangon**, or Rangoon, is its capital and a major port city. A harsh military government rules Myanmar. This government abuses

 Interactive Reader and Study Guide

Section 3, *continued*

human rights—rights that all people deserve. One woman, Aung San Suu Kyi, is working to reform Myanmar so people have more rights. Suu Kyi and others have been arrested, however. Because of Myanmar's poor record on human rights, many countries will not trade with it. As a result, Myanmar's economy has suffered.

Thailand, once called Siam, has a strong economy. Its capital and largest city is **Bangkok**, a city famous for its **klongs**, or canals. Bangkok's klongs are used for trade and travel, and to drain floodwater. Thailand has a democratically elected government and rich resources. These factors have helped its economy grow.

> Why won't some countries trade with Myanmar?
> _____________________
> _____________________

THE COUNTRIES OF INDOCHINA

After decades of war, the countries of Indochina—Cambodia, Laos, and Vietnam—are working hard to improve their economies.

The capital and chief city of Cambodia, **Phnom Penh**, is a center of trade. However, years of war have left the country with little industry. Although farming has improved, many landmines still remain buried.

> How has war affected the countries of Indochina?
> _____________________
> _____________________

Laos is the area's poorest country. It is landlocked, with few roads, no railroads, and little electricity. Most people are subsistence farmers, just growing enough food to support their families.

Vietnam's main cities include its capital in the north, **Hanoi**, and Ho Chi Minh City in the south. Although still a Communist country, Vietnam's government has been allowing more economic freedoms. Industry and services are growing, but most people still farm.

CHALLENGE ACTIVITY

Critical Thinking: Summarizing Summarize the information about the countries of Mainland Southeast Asia, using a graphic organizer.

Southeast Asia

Section 4

MAIN IDEAS

1. The area today has rich resources and growing cities but faces challenges.
2. Malaysia and its neighbors have strong economies but differ in many ways.
3. Indonesia is big and diverse with a growing economy, and East Timor is small and poor.
4. The Philippines has less ethnic diversity, and its economy is improving.

Key Terms and Places

kampong village or city district with traditional houses built on stilts; slums around cities

Jakarta capital of Indonesia

Kuala Lumpur Malaysia's capital and a cultural and economic center

free ports ports that place few if any taxes on goods

sultan supreme ruler of a Muslim country

Java Indonesia's main island

Manila capital of the Philippines

Section Summary

THE AREA TODAY

The countries of Southeast Asia are Malaysia, Singapore, Brunei, Indonesia, East Timor, and the Philippines. Like the mainland countries, these island countries face challenges, too, such as ethnic conflicts, poverty, and environmental problems.

As on the mainland, many people in Island Southeast Asia live in rural areas. They fish or farm. Rice is the main crop. Also like the mainland, many people here are moving to cities for work. In some areas, people live in **kampongs**, places with traditional houses on stilts. The term *kampong* also refers to slums around **Jakarta**, Indonesia's capital, and around other cities of Island Southeast Asia.

> Underline the challenges facing Island Southeast Asia.

> Circle the main crop of this area.

  Interactive Reader and Study Guide

MALAYSIA AND ITS NEIGHBORS

Malaysia has two parts. One is on the Malay Peninsula, where most Malaysians live. Its capital, **Kuala Lumpur**, is there as well. The other part is on Borneo. Malaysia is ethnically diverse, with a strong economy. It has a constitutional monarchy, with a prime minister and elected legislature.

Singapore is located on a tiny island that lies on a major shipping route. It is one of the world's busiest **free ports**—ports with few or no taxes on goods. Singapore is modern, wealthy, and clean. It has a strict government. Brunei is on the island of Borneo. A **sultan**, the supreme ruler of a Muslim country, governs this tiny country. As a result of oil and gas deposits, Brunei has grown wealthy.

> **How does Singapore's location help its economy?**
> ___________________________
> ___________________________

INDONESIA AND EAST TIMOR

Indonesia is the world's largest archipelago. It has the world's largest Muslim population and is ethnically diverse. **Java** is Indonesia's main island. More than half its people live there. Indonesia has rich resources, such as rubber, oil, gas, and timber. Its economy is growing, but many people are poor. Religious and ethnic conflicts have led to violence.

After breaking free from Indonesia, years of fighting have left the tiny country of East Timor poor. Most people there farm.

> **What are some of Indonesia's resources?**
> ___________________________
> ___________________________

THE PHILIPPINES

The Philippines has less ethnic diversity than other island countries. **Manila** is its capital. Although some Filipinos are rich, most are poor farmers who do not own any land. The country is mainly Roman Catholic, but some areas are mostly Muslim and are seeking independence.

CHALLENGE ACTIVITY

Critical Thinking: Drawing Inferences Choose an island country of Southeast Asia and write an essay that explains what steps could be taken to improve its economy.

The Pacific World

CHAPTER SUMMARY

	Physical Geography	History/Culture	Area Today
Australia and New Zealand	• Australia: flat, hot, dry in west; low mountains, valleys, river system in east	• similar history and culture	• Most people are of British descent.
The Pacific Islands	• regions: Micronesia, Melanesia, Polynesia	• Settled at least 35,000 years ago • 1500s: Europeans came	• developing economies; fishing, tourism, agriculture
Antarctica	• frozen continent at South Pole; ice covers 98 percent	• 1775: James Cook first sighted the Antarctic Peninsula	• only continent without a permanent human population

COMPREHENSION AND CRITICAL THINKING

Use information from the graphic organizer to answer the following questions.

1. Compare and Contrast How are Australia and New Zealand alike and different?

____

____

____

2. Make Judgments Should countries be allowed to claim parts of Antarctica? Why or why not?

____

____

____

The Pacific World

Section 1

MAIN IDEAS

1. The physical geography of Australia and New Zealand is diverse and unusual.
2. Native peoples and British settlers shaped the history of Australia and New Zealand.
3. Australia and New Zealand today are wealthy and culturally diverse countries.

Key Terms and Places

Great Barrier Reef largest coral reef in the world, off Australia's northeastern coast

coral reef a collection of rocky material found in shallow, tropical waters

Aborigines first humans to live in Australia

Maori New Zealand's first settlers

Outback Australia's interior

Section Summary

PHYSICAL GEOGRAPHY

Australia has wide, flat stretches of dry land. Low mountains, valleys, and a major river system make up the eastern past of the country. The **Great Barrier Reef**, the world's largest **coral reef**, is located off of Australia's northeastern coast.

New Zealand is made up of North Island and South Island. A large mountain range called the Southern Alps is a key feature of South Island. New Zealand also has green hills, volcanoes, hot springs, dense forests, deep lakes, and fertile plains.

Most of Australia has warm, dry desert and steppe climates. The coastal areas are milder and wetter. New Zealand has a marine climate with plenty of rainfall and mild temperatures. Native animals include Australia's kangaroo and koala and New Zealand's kiwi, a flightless bird. Australia has many valuable mineral resources, and farms raise wheat, cotton, and sheep despite the poor soil. New Zealand has few mineral resources, but plenty of rich soil.

> Underline the two islands that make up New Zealand.

> Circle the region's native animals.

HISTORY

Aborigines, the first humans in Australia, came from Southeast Asia more than 40,000 years ago. Early Aborigines were nomads who gathered plants, hunted animals, and fished. The **Maori**, New Zealand's first settlers, came from other Pacific islands about 1,200 years ago. They hunted, fished, and farmed. Captain James Cook visited New Zealand in 1769. British settlers began arriving in Australia and New Zealand in the late 1700s and early 1800s. Australia and New Zealand gained independence in the early 1900s. Both countries are in the British Commonwealth and close allies of the United Kingdom.

> Who were the first humans to settle in Australia and New Zealand?
>
> _______________________
>
> _______________________

AUSTRALIA AND NEW ZEALAND TODAY

Most people in both countries are of British ancestry. Sydney and Melbourne are Australia's two largest cities. Auckland is New Zealand's largest city. Both countries produce wool, meat, and dairy products. Mining is important throughout the **Outback**. Other industries include steel, heavy machines, and computers. Manufacturing, banking, and tourism are important in New Zealand.

> Circle the name of New Zealand's largest city.

Today, Australia and New Zealand face the challenge of improving the political and economic status of their native populations.

CHALLENGE ACTIVITY

Critical Thinking: Making Judgments Why do you think the Aboriginal and Maori populations of Australia and New Zealand declined after British settlers arrived? Explain your answer in a short paragraph.

The Pacific World

Section 2

MAIN IDEAS

1. Unique physical features, tropical climates, and limited resources shape the physical geography of the Pacific Islands.
2. Native customs and contact with the western world have influenced the history and culture of the Pacific Islands.
3. Pacific Islanders today are working to improve their economies and protect the environment.

Key Terms and Places

Micronesia region of Pacific Islands located east of the Philippines

Melanesia region of Pacific Islands stretching from New Guinea to Fiji

Polynesia largest region of Pacific Islands, east of Melanesia

atoll small, ring-shaped coral island surrounding a lagoon

territory an area that is under the control of another government

Section Summary

PHYSICAL GEOGRAPHY

There are three regions of islands in the Pacific. **Micronesia**, consisting of about 2,000 small islands, is east of the Philippines. **Melanesia**, the most heavily populated region, stretches from New Guinea to Fiji. **Polynesia**, the largest region, is located east of Melanesia.

There are two kinds of islands in the Pacific: high islands and low islands. High islands are formed from volcanoes or continental rock. They tend to be mountainous and rocky. They have dense forests, rich soil, and many mineral resources. Low islands are much smaller; they have thin soil, little vegetation, few resources, and low elevations. Many low islands are **atolls**, small ring-shaped coral islands surrounding lagoons.

Most high and low islands have a humid tropical climate. Temperatures are warm and rain falls all year.

> **What three regions make up the Pacific Islands?**
>
> __________________
> __________________
> __________________

> **Circle the type of climate that most high and low islands have.**

Section 2, *continued*

HISTORY AND CULTURE

People began settling the Pacific more than 40,000 years ago. They arrived in Melanesia first. Polynesia was the last region to be settled. Europeans first encountered the Pacific Islands in the 1500s. John Cook, a captain in the British Navy visited all the main regions in the 1700s. By the late 1800s, Britain, Spain, France, and other European nations gained control of most of the islands. When the United States defeated Spain in the Spanish-American War, it took Guam as a **territory**, which is an area that is under the authority of another government. After World War I, Japan gained control of many islands. After World War II, the United Nations placed some islands under the control of the United States and its Allies.

Nearly 8 million people of many cultures and ethnic groups live in the Pacific Islands. Most are descended from the original settlers. Some are ethnic Europeans and Asians. Most islanders are now Christian. Many, however, continue to practice traditional customs ranging from building constructions to art styles and various ceremonies.

> Circle the name of the Pacific Island region that was first to be settled.

> What country took Guam as a territory after the Spanish-American war?
> ___________________________

> About how many people live in the Pacific Islands?
> ___________________________

THE PACIFIC ISLANDS TODAY

The Pacific Islands face important challenges today. They are trying to build stronger economies through tourism, agriculture, and fishing. Some countries, including Papua New Guinea, export gold, copper, and oil. The islands also must cope with the potentially damaging effects of past nuclear testing in the area and global warming.

> Underline three ways the Pacific Islands are trying to improve their economies.

CHALLENGE ACTIVITY

Critical Thinking: Drawing Conclusions Which island countries probably have stronger economies: those occupying high islands or those occupying low islands? Support your answer using details from the summary.

The Pacific World

Section 3

MAIN IDEAS

1. Freezing temperatures, ice, and snow dominate Antarctica's physical geography.
2. Explorations in the 1800s and 1900s led to Antarctica's use for scientific research.
3. Research and protecting the environment are key issues in Antarctica today.

Key Terms and Places

ice shelf ledge of ice that extends over the water

icebergs floating masses of ice that have broken off a glacier

Antarctic Peninsula peninsula that extends north of the Antarctic Circle

polar desert a high-latitude region that receives very little precipitation

ozone layer layer of Earth's atmosphere that protects living things from the harmful effects of the sun's ultraviolet rays

Section Summary

PHYSICAL GEOGRAPHY

Ice covers more than 98 percent of Antarctica. Ice flows slowly toward the coasts of Antarctica. At the coast it forms a ledge over the water called an **ice shelf**. Antarctica's largest ice shelf is the Ross Ice Shelf. Floating masses of ice that break off from ice shelves are called **icebergs**. The **Antarctic Peninsula** is on the western side of the continent.

Antarctica has a freezing ice cap climate. It is a **polar desert**, a high-latitude region that receives little precipitation. It is bitterly cold during the dark winter. Summer temperatures reach near freezing.

Tundra plant life grows in ice-free areas. Insects are the only animals on the land. Penguins, seals, and whales live in the nearby waters. Antarctica has many mineral resources.

> **How is an iceberg related to an ice shelf?**
>
> _______________________
> _______________________
> _______________________

     Interactive Reader and Study Guide

EARLY EXPLORATIONS

Antarctica was first sighted in 1775. European explorations investigated Antarctica throughout the 1800s. A team of Norwegian explorers became the first people to reach the South Pole in 1911.

Several countries have claimed parts of Antarctica. In 1959, several countries signed the Antarctic Treaty, which bans military activity on the continent and sets it aside for scientific research.

> **What are the terms of the Antarctic Treaty of 1959?**
> _______________________
> _______________________
> _______________________

ANTARCTICA TODAY

Antarctica is the only continent with no permanent human population. Scientists use Antarctica to conduct research and study the environment. Several countries, including the United States, have bases there.

Antarctic research includes studies of plant and animal life, analyzing weather conditions and patterns, and examining issues affecting Earth's environment. Scientists are concerned about the thinning of the **ozone layer**—the layer of Earth's atmosphere that protects living things from the harmful effects of the sun's ultraviolet rays.

Tourism, oil spills, and mining are real and potential threats to Antarctica's environment. An agreement reached in 1991 bans mining and drilling and limits tourism.

> **Underline three activities that threaten Antarctica's environment.**

CHALLENGE ACTIVITY

Critical Thinking: Making Judgments What do you think is the greatest threat to Antarctica's environment? Why? Explain your answer.